STUDY GUIDE

for the Seventh Edition of

Fire and Emergency Services Instructor

Project Manager
Beth Ann Fulgenzi

Edited By
Melissa Noakes
Patrice Barnett

Technical Review
Frederick M. Stowell

Published by
Fire Protection Publications
Oklahoma State University
Stillwater, Oklahoma

The International Fire Service Training Association

The International Fire Service Training Association (IFSTA) was established in 1934 as a "non-profit educational association of fire fighting personnel who are dedicated to upgrading fire fighting techniques and safety through training." To carry out the mission of IFSTA, Fire Protection Publications was established as an entity of Oklahoma State University. Fire Protection Publications' primary function is to publish and disseminate training texts as proposed and validated by IFSTA. As a secondary function, Fire Protection Publications researches, acquires, produces, and markets high-quality learning and teaching aids consistent with IFSTA's mission.

NOTICE: The questions in this study guide are taken from the information presented in the seventh edition of ***Fire and Emergency Services Instructor***, an IFSTA-validated manual. The questions are *not validated test questions and are not intended to be duplicated or used for certification or promotional examinations;* this guide is intended to be used as a tool for studying the information presented in ***Fire and Emergency Services Instructor***.

ISBN 0-87939-272-X

First Edition
First Printing, March 2006

Printed in the United States of America 10 9 8 7 6 5 4 3 2

If you need additional information concerning our organization or assistance with manual orders, contact:
Customer Service, Fire Protection Publications, Oklahoma State University
930 N. Willis, Stillwater, OK 74078-8045
1-800-654-4055 FAX: 405-744-8204

For assistance with training materials, to recommend material for inclusion in an IFSTA manual, or to ask questions on manual content, contact:
Editorial Department, Fire Protection Publications, Oklahoma State University
930 N. Willis, Stillwater, OK 74078-8045
405-744-5723 FAX: 405-707-0024 E-mail: editors@osufpp.org

Table of Contents

Answers to Chapter Questions

Preface

This study guide is designed to help the reader understand and remember the material presented in **Fire and Emergency Services Instructor**, seventh edition. It identifies important information and concepts from each chapter and provides questions to help the reader study and retain this information. In addition, the study guide serves as an excellent resource for individuals preparing for certification or promotional examinations. The questions in this guide are designed to help you remember information and to make you think—they are *not* intended to trick or mislead you.

Chapter 1 Challenges of Fire and Emergency Services Instruction

Multiple Choice

A. Write the correct letters on the blanks.

_____ 1. Which of the following is a solution to the challenge of ensuring safe training environments?

A. Do not train at remote sites.
B. Limit the time devoted to practical evolutions.
C. Be familiar with safety regulations as they apply to training.
D. Let only advanced students perform practical training evolutions.

_____ 2. Which of the following challenges is met by taking advantage of continuing education?

A. Multiple priorities
B. Organizational apathy
C. Professional development
D. Qualified instructor recruitment

_____ 3. Which of the following challenges is met by ensuring that external customers are provided with the same high-quality training that internal customers receive?

A. Cooperative relationships
B. Organizational promotion
C. Management directives
D. Funds and resources

_____ 4. An instructor's primary obligation is to whom?

A. Students
B. The public
C. The profession
D. The organization

_____ 5. To which of the following does the instructor have the obligation of providing the new recruit with a link to the heritage and tradition of the emergency services?

A. Students
B. Themselves
C. Profession
D. Organization

_____ 6. Which of the following is NOT a characteristic of effective instructors?
A. Desire to teach
B. Enthusiasm
C. Ability to be subjective
D. Subject and teaching competencies

_____ 7. Which of the following is the ability to understand the feelings and attitudes of another person?
A. Sympathy
B. Empathy
C. Objectivity
D. Subjectivity

_____ 8. Which of the following is NOT an effective leadership quality for instructors?
A. Arrogance
B. Expertise
C. Consistency
D. Trustworthiness

_____ 9. Which of the following statements about the instructor as a role model is TRUE?
A. Adult students are not influenced by role models.
B. Being a role model is not the responsibility of the instructor.
C. An instructor's influence is limited to the classroom.
D. An instructor's influence goes beyond the classroom.

_____ 10. Which of the following is a guideline for managing diversity issues?
A. Show favoritism to a single student or group.
B. Adapt teaching methods to the members of the audience.
C. Openly criticize a student in the presence of others.
D. Permit audience members to use inappropriate language.

_____ 11. Which of the following is a guideline for audience awareness?
A. Patronize or talk down to the audience.
B. Use job-specific jargon when possible.
C. Present all of the material in one session without rest breaks.
D. Permit questions from the audience either during or at the end of presentations.

_____ 12. Which of the following is a guideline for the use of instructional facilities, props, and acquired structures?

A. Let visitors and spectators on the training grounds.
B. Perform training in inclement weather if already scheduled.
C. Modify commercially produced props whenever needed.
D. Have a designated incident safety officer (ISO) present during all high-hazard training activities.

_____ 13. Which of the following is NOT a guideline for dealing with legal requirements and accommodations?

A. Ensure that accessibility is provided for instructional classrooms and facilities.
B. Provide a teaching environment that ensures the best possible learning experience for all students.
C. Provide appropriate handouts or training aids for students who may be visually or audibly impaired.
D. Assume that all students are physically, mentally, and emotionally similar and able to learn in the same environment.

_____ 14. Which of the following is defined as the use of language that may be insulting, demeaning, or scornful of others?

A. Sarcasm
B. Sincerity
C. Bluffing
D. Positive humor

_____ 15. Which of the instructor characteristics refers to providing the same learning opportunities for all students and evaluating performance against an established objective standard and not against a subjective set of results?

A. Empathy
B. Sincerity
C. Fairness
D. Personal integrity

True / False

B. Write the correct letters on the blanks.

_____ 1. An instructor has an obligation to continue professional development through acquiring new knowledge and improving skills.

A. True
B. False

_____ 2. If instructors do not know the answer to a question, it is appropriate for them to bluff their way through quickly and move on to the next subject.

A. True
B. False

_____ 3. Offensive or inappropriate humor, jokes, language, and stories that belittle or degrade a person or group of persons have no place in any training session.
A. True
B. False

_____ 4. An instructor can reasonably expect all students to act, think, respond, and learn the same.
A. True
B. False

_____ 5. When performing analyses it is acceptable to depend on one source for information if that source is a product vendor or manufacturer.
A. True
B. False

_____ 6. When identifying legal requirements and accommodations the instructor should provide appropriate handouts or training aids for students who may be visually or audibly impaired.
A. True
B. False

_____ 7. When recognizing audience characteristics the instructor should ask questions of audience members to ensure that they understand the information.
A. True
B. False

_____ 8. Training sessions do not need to use the adopted accountability system and the Incident Command System.
A. True
B. False

_____ 9. EMS personnel must be available during live-fire and technical-rescue training exercises.
A. True
B. False

_____ 10. One solution to the challenge of adequate funding and resources is to seek funding from new sources such as grants, business and industrial sponsors/partnerships, and donations.
A. True
B. False

_____ 11. Intimidation can be an effective tool for instructors if students are not putting forth enough effort in the classroom.

A. True
B. False

_____ 12. The use of humor can add emphasis to and create interest in a subject.

A. True
B. False

_____ 13. Instructors have an obligation to students to teach safe operational practices and safety-related topics.

A. True
B. False

_____ 14. Instructors have an obligation to provide feedback to students and supervisors.

A. True
B. False

_____ 15. Instructors have an obligation to present a professional image both on and off duty.

A. True
B. False

List

C. Write effective characteristics of instructors on the lines provided.

1. ______________________________
2. ______________________________
3. ______________________________
4. ______________________________
5. ______________________________
6. ______________________________
7. ______________________________
8. ______________________________
9. ______________________________

10. ____

11. ____

12. ____

Chapter 2 Safety and the Training Function

Multiple Choice

A. Write the correct letters on the blanks.

_____ 1. What happens as training becomes more realistic?

A. The benefit to the student decreases and potential risk increases.
B. The benefit to the student increases and potential risk increases.
C. The benefit to the student decreases and potential risk decreases.
D. The benefit to the student increases and potential risk decreases.

_____ 2. What is the primary strategic goal of all incident action plans?

A. Life safety
B. Expense reduction
C. Positive feedback
D. Personnel management

_____ 3. What does a hazard/risk analysis identify?

A. Potential safety threats
B. Number of actual incidents
C. Number of emergency calls
D. Costs of responding to incidents

_____ 4. Who has the ultimate responsibility for safety-related issues during either an emergency incident or training scenario?

A. Instructor
B. Incident commander
C. Incident safety officer
D. Health safety officer

_____ 5. Who is the primary role model for safety?

A. Peers
B. Instructor
C. Incident commander
D. Incident safety officer

_____ 6. Which of the following is NOT an item of information that students need to know in case of an accident?

A. Whether training stops or proceeds
B. The name and rank of injured individuals
C. What steps to take to help the injured person
D. What signals are given when an accident occurs

_____ 7. Which of the following statements concerning accidents is TRUE?

A. Most accidents cannot be prevented from occurring.
B. Almost one-half of all accidents could be prevented.
C. The majority of accidents are predictable and preventable.
D. Accidents cannot happen to individuals following safety precautions.

_____ 8. What are mitigation activities?

A. Activities that decrease legal liability
B. Activities that increase the probability of an accident
C. Activities that totally eliminate the effects of a hazard
D. Activities that prevent or minimize the effects of a hazard

_____ 9. Which of the following agencies has the responsibility of setting and enforcing workplace safety and health standards and has the authority to issue citations and fines?

A. Department of Homeland Security (DHS)
B. Federal Emergency Management Agency (FEMA)
C. Occupational Safety and Health Administration (OSHA)
D. National Institute for Occupational Safety and Health (NIOSH)

_____ 10. Which of the following organizations develops minimum safety standards and guidelines relating to live-fire training evolutions and high-hazard training?

A. Underwriters Laboratories, Inc. (UL)
B. National Fire Protection Association (NFPA)
C. American National Standards Institute (ANSI)
D. American Society for Testing and Materials (ASTM)

True / False

B. Write the correct letters on the blanks.

_____ 1. An accident is a sequence of unplanned or uncontrolled events that produces unintended injuries, deaths, or property damage.
A. True
B. False

_____ 2. The instructor's responsibility is to provide realistic training without regard to the level of safety.
A. True
B. False

_____ 3. Injuries and fatalities can occur at the incident scene, during training, during work shifts at the station, and when responding to and returning from incidents.
A. True
B. False

_____ 4. One recommendation for casualty prevention is to avoid the use of live-burn evolutions in any structure types.
A. True
B. False

_____ 5. One method by which a jurisdiction can reduce risk is to establish and adhere to the National Incident Management System (NIMS) Incident Command System.
A. True
B. False

_____ 6. The authority having jurisdiction should establish an inspection time schedule for facilities and props based on industry practice, manufacturer's recommendations, and local needs.
A. True
B. False

_____ 7. The presence of an incident safety officer relieves instructors who are present of the obligation to monitor the training with safety in mind.
A. True
B. False

_____ 8. The health and safety officer is responsible for teaching safety-related topics and that responsibility is never delegated to an Instructor I.

A. True
B. False

_____ 9. At either an emergency incident or training scenario where activities are judged by the incident safety officer to be unsafe or involve an imminent hazard, the incident safety officer has the authority to alter, suspend, or terminate those actions.

A. True
B. False

_____ 10. When conducting practical training evolutions, instructors do not need to appoint an incident safety officer.

A. True
B. False

_____ 11. Accidents are usually the result of circumstances that are outside of any individual's control.

A. True
B. False

_____ 12. Human factors that can contribute to accidents include improper attitude, lack of knowledge or skill, and physical limitations.

A. True
B. False

_____ 13. One purpose of an accident investigation is to determine the change or deviation that caused the accident.

A. True
B. False

_____ 14. The instructor is generally responsible for accident investigation.

A. True
B. False

_____ 15. The first step to safe training programs is preventing accidents, which means preventing the loss of personnel, property, money, and time.

A. True
B. False

List

C. **Write the four places where injuries and fatalities can occur on the lines provided.**

1. ______________________________
2. ______________________________
3. ______________________________
4. ______________________________

Terms

D. **Describe the following terms on the lines provided.**

1. Task analysis ______________________________

____

____

2. Hazard/risk analysis ______________________________

____

____

3. Incident Action Plan ______________________________

____

____

____

Chapter 3 Legal and Ethical Considerations

Multiple Choice

A. Write the correct letters on the blanks.

_____ 1. Which of the following refers to an instrument that provides direction but does not have the force of law?
A. Code
B. Guide
C. Standard
D. Regulation

_____ 2. Which of the following refers to authoritative rule dealing with details of procedures or a rule or order having the force of law that is issued by an executive authority of government?
A. Code
B. Guide
C. Standard
D. Regulation

_____ 3. Which of the following refers to a body of law established either by legislative or administrative agencies with rule-making authority that is designed to regulate, within its scope, the topic to which it relates?
A. Code
B. Guide
C. Standard
D. Regulation

_____ 4. Which of the following type of law is usually the result of a legal precedent?
A. Judiciary law
B. Legislative law
C. Administrative law
D. Regulatory law

_____ 5. Which of the following type of law is created by regulatory agencies?
A. Judiciary law
B. Legislative law
C. Administrative law
D. Regulatory law

_____ 6. Which of the following type of law is made by federal, state/provincial, and local governments to address the needs of citizens?
A. Judiciary law
B. Legislative law
C. Administrative law
D. Regulatory law

_____ 7. Which of the following laws was passed in 1964 and has the purpose of providing stronger protection for rights guaranteed by the U.S. Constitution?
A. Civil Rights Act
B. Privacy Act
C. Buckley Amendment
D. Americans with Disabilities Act

_____ 8. Which of the following laws prohibits certain questions of job applicants?
A. Civil Rights Act
B. Privacy Act
C. Title VII of the Civil Rights Act
D. Americans with Disabilities Act

_____ 9. Which of the following laws created the equal employment opportunity (EEO) law?
A. Civil Rights Act
B. Privacy Act
C. Title VII of the Civil Rights Act
D. Americans with Disabilities Act

_____ 10. Which of the following laws restricts access to personal information such as personnel files and student grades?
A. Civil Rights Act
B. Affirmative action policies
C. Buckley Amendment
D. Americans with Disabilities Act

_____ 11. Which of the following types of records must be kept confidential?
A. Incident reports
B. Official meeting minutes
C. Budget and expense reports
D. Individual training records

_____ 12. Which of the following refers to liability that is placed on the employer for the acts and omissions of employees during the normal course of their employment?

A. Secondary liability
B. Vicarious liability
C. Employer liability
D. Tertiary liability

_____ 13. Which of the following is NOT a precaution that instructors can take to reduce liability?

A. Check equipment regularly for safe operating conditions.
B. Maintain written objectives, and document each training session.
C. Have students sign of waiver releasing the instructor from liability in case of injury.
D. Do NOT exceed individual skill level when training students or working with other instructors.

_____ 14. Which of the following governs the majority of copyright law in the United States?

A. Local law
B. State statute
C. Federal statute
D. International law

_____ 15. Which of the following would NOT be considered fair use when preparing to teach a class?

A. Chapter from a book
B. Short story, essay, or poem
C. Cartoon from a newspaper
D. Monthly issue of a periodical

_____ 16. Which of the following agencies regulates and controls activities that may have a negative effect on the nation's water, atmosphere, and soil?

A. Department of Homeland Security
B. Environmental Protection Agency
C. National Environmental Cooperative
D. Environmental Control and Regulatory Commission

_____ 17. Which of the following may prohibit or control the intentional release of hydrocarbons into the atmosphere from a live-fire exercise?

A. Federal statutes
B. Regional statutes
C. Neighborhood guidelines
D. Local open-burning ordinances

_____ 18. Which of the following statements about soil and training evolutions is TRUE?

A. Water runoff is unlikely to contaminate soil.
B. Contaminated soil is too difficult to remove and should be left alone.
C. Soil may become contaminated and have to be removed.
D. Surface contamination of soil does not need to be removed.

_____ 19. Which of the following statements concerning fuels for live-fire training exercises is TRUE?

A. Never use props equipped with natural gas simulators.
B. Spread of contaminants does not need to be monitored.
C. Fuels with the lowest cost should be used for training exercises.
D. Fuels that emit the minimum amount of contaminants should be used.

_____ 20. Which of the following is the primary base for the origin of personal ethics?

A. Peers
B. Family
C. Organized religion
D. Educational institutions

True / False

B. Write the correct letters on the blanks.

_____ 1. Instructors and the organizations they work for can be held legally responsible for their actions or lack of actions.

A. True
B. False

_____ 2. Consensus standards, when legally adopted, establish minimum criteria for student performance and evaluation during training.

A. True
B. False

_____ 3. The most common standards in the fire and emergency services are standards developed by the Occupational Safety and Health Administration.

A. True
B. False

_____ 4. The three types of law are legislative, administrative, and political.

A. True
B. False

_____ 5. Local laws must meet or exceed the federal law.
A. True
B. False

_____ 6. Affirmative action policies establish employment programs required by federal statutes and regulations designed to correct past and current discriminatory practices in hiring members of underutilized and minority groups.
A. True
B. False

_____ 7. Ordinances are created by local governments and have less force than statutory laws.
A. True
B. False

_____ 8. Daily attendance records for all personnel are primarily maintained to provide data for the distribution of payrolls and benefits.
A. True
B. False

_____ 9. The length of time that records must be retained by the organization is either three years or five years, depending upon the type of record.
A. True
B. False

_____ 10. Official meeting minutes and any other notes that are made as part of a meeting are considered confidential.
A. True
B. False

_____ 11. Test scores and personal data are considered privileged information and are available only to management and designated personnel with authorization and a specific need to know.
A. True
B. False

_____ 12. Instructors may be considered negligent for teaching a topic they are unqualified to teach.
A. True
B. False

_____ 13. Ensuring that the training is as realistic as possible should be considered more important than providing a safe training environment.
A. True
B. False

_____ 14. Instructors cannot be expected to predict that certain hazardous conditions will be present during training.
A. True
B. False

_____ 15. One precaution that instructors can take to reduce liability is to never leave students unattended while they are practicing potentially dangerous skills.
A. True
B. False

_____ 16. One step that instructors can take to prevent or minimize personal liability in the event of an accident is to document all issues of discrepancy, complaint, and injury accurately.
A. True
B. False

_____ 17. If an individual is photographed, that individual has no control over how or where the photograph may be used.
A. True
B. False

_____ 18. Students should have the freedom to hold and express an opinion that may be in opposition to the organization or instructor.
A. True
B. False

_____ 19. A written code of ethics establishes a framework for professional behavior and strengthens the organization's ethical climate.
A. True
B. False

_____ 20. When resolving an ethical dilemma, the individual should look at the facts subjectively.
A. True
B. False

List

C. **Write four ways that instructors can reduce the potential for liability and legal action against themselves and their organizations on the lines provided.**

1. ______________________________

____

2. ______________________________

____

3. ______________________________

____

4. ______________________________

____

Terms

D. **Describe the following terms on the lines provided.**

1. Fair use doctrine of the Copyright Act ______________________________

____

____

2. Invasion of privacy ______________________________

____

____

Chapter 4 Effective Interpersonal Communication

Multiple Choice

A. Write the correct letters on the blanks.

_____ 1. Which of the following basic elements of interpersonal communication originates a message by encoding or turning thoughts and mental images into words?

A. Channel
B. Sender
C. Receiver
D. Message

_____ 2. Which of the following elements of interpersonal communication is NOT essential for effective interpersonal communication to occur?

A. Medium
B. Receiver
C. Feedback
D. Interference

_____ 3. Which of the following may prevent the receiver from fully receiving a message?

A. Medium
B. Channel
C. Feedback
D. Interference

_____ 4. Which of the following will cause the message's effect to be obvious to the sender by the auditory, visual, gestural, or tactile response of the receiver?

A. Feedback
B. Interference
C. Perception
D. Context

_____ 5. Which of the following purposes of interpersonal communication aims to control, direct, or manipulate behavior?

A. Learning
B. Relating
C. Influencing
D. Entertaining

_____ 6. Which of the following components of the listening process completes the communication process and means an exchange of roles has occurred?

A. Attending
B. Understanding
C. Remembering
D. Responding

_____ 7. Which of the following components of the listening process involves decoding the message and assigning meaning to it?

A. Attending
B. Understanding
C. Remembering
D. Responding

_____ 8. Which of the following terms refers to a generalization that may not be verifiable without additional data?

A. Facts
B. Opinions
C. Statements
D. Feedback

_____ 9. Which of the following is the preferred approach to radio communication in the fire service?

A. Agency code
B. 10-codes
C. Command text
D. Clear-text (plain English)

_____ 10. Which of the following creates the perception of a hostile work environment that is a major impediment to successful leadership?

A. Use of humor
B. Acknowledging differences
C. Use of stereotypes
D. Use of generalizations

_____ 11. When should you avoid addressing a problem?

A. While in a hurry
B. While in front of students
C. While in front of supervisors
D. While angry or emotional

_____ 12. Which of the following is NOT a main element of kinesics?

A. Gestures
B. Eye contact
C. Facial expressions
D. Inflection

_____ 13. Which of the following is a guideline for verbal skills improvement?
A. Use generalizations.
B. Engage in dual perspective.
C. Apply personal feelings to another person.
D. Blame others for personal feelings and thoughts.

_____ 14. Which of the following is a guideline for improving nonverbal communication?
A. Avoid direct eye contact.
B. Avoid acknowledging cultural differences.
C. Eliminate inflection from the communication process.
D. Learn to match the facial expression to the message.

_____ 15. Which of the following terms refers to restating the message in different words but keeping the same meaning?
A. Repeating
B. Responding
C. Interpreting
D. Paraphrasing

True / False

B. Write the correct letters on the blanks.

_____ 1. The receiver's frame of reference depends on education, cultural background, perception, attitude, and context.
A. True
B. False

_____ 2. The preferred approach to radio communication in the fire service is 10-codes.
A. True
B. False

_____ 3. Interference is an essential element for interpersonal communication to occur.
A. True
B. False

_____ 4. The meaning or symbolism that people place on words depends on the cultural background of people.
A. True
B. False

_____ 5. Instructors should use technical language and fire service jargon when speaking with individuals from outside the profession.
A. True
B. False

_____ 6. Slurs, innuendos, name calling, and inappropriate jokes and comments are not acceptable in the fire and emergency services profession.
A. True
B. False

_____ 7. Listening is an active process that includes attending, understanding, remembering, evaluating, and responding to the speaker.
A. True
B. False

_____ 8. The best way to improve listening skills is to practice them.
A. True
B. False

_____ 9. Individuals should NOT take notes during formal speeches since this can distract from the message.
A. True
B. False

_____ 10. To improve listening skills individuals should respond to the speaker by asking questions or paraphrasing what has been said.
A. True
B. False

List

C. Write the five essential basic elements of interpersonal communication on the lines provided.

1. ____________________
2. ____________________
3. ____________________
4. ____________________
5. ____________________

Describe

D. Describe the following purposes of interpersonal communication on the lines provided.

1. Learning ____________________

____

2. Relating ____________________

____

3. Influencing ____________________

____

4. Playing ____________________

____

5. Helping ____________________

____

Chapter 5 Instructional Facilities and Props

Multiple Choice

A. Write the correct letters on the blanks.

_____ 1. Which of the following NFPA standards sets forth requirements for live-fire training?
- A. NFPA 1021
- B. NFPA 1403
- C. NFPA 1520
- D. NFPA 1605

_____ 2. Which of the following is NOT an infrastructure requirement for the training location?
- A. Easily accessible
- B. Large enough for potential expansion
- C. Nearby other local government facilities
- D. Remote from other occupancies that may be affected by live burns

_____ 3. Which of the following factors does NOT affect water supply required for training operations?
- A. Number of attack and backup hoselines used
- B. Need for potable (suitable for drinking) water
- C. Need to supply water for nearby residences for general use
- D. Need to supply water for other types of exercises that may take place at the site

_____ 4. Which of the following refers to a multistory, multipurpose structure that may be open or enclosed and should be optimally six stories in height?
- A. Drill tower
- B. Acquired structure
- C. Live-fire buildings
- D. Smoke buildings

_____ 5. Which of the following structures must be able to withstand high temperatures created by either fueled props or class A fuel loads?
- A. Drill tower
- B. Acquired structure
- C. Live-fire buildings
- D. Smoke buildings

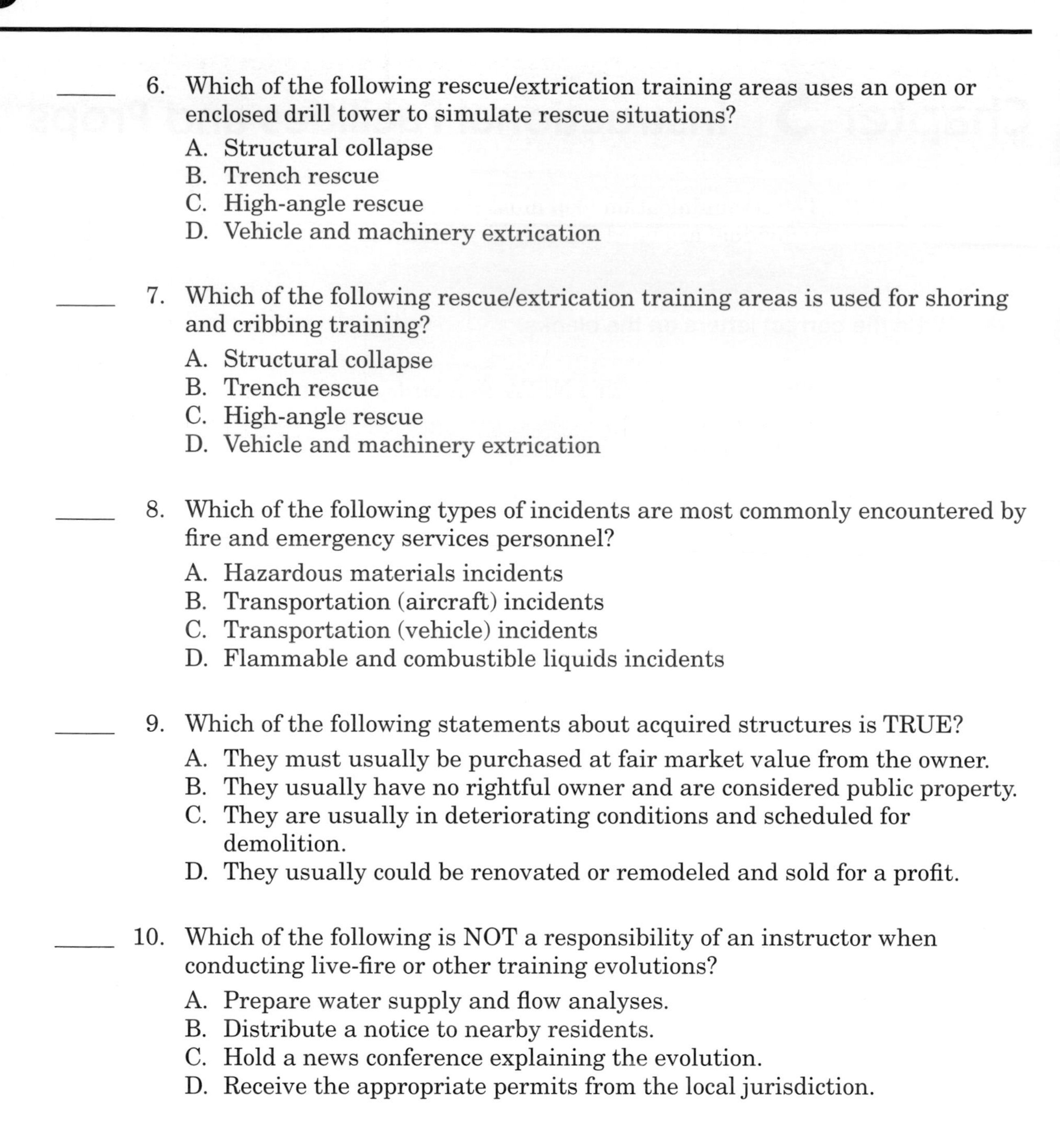

_____ 6. Which of the following rescue/extrication training areas uses an open or enclosed drill tower to simulate rescue situations?

A. Structural collapse
B. Trench rescue
C. High-angle rescue
D. Vehicle and machinery extrication

_____ 7. Which of the following rescue/extrication training areas is used for shoring and cribbing training?

A. Structural collapse
B. Trench rescue
C. High-angle rescue
D. Vehicle and machinery extrication

_____ 8. Which of the following types of incidents are most commonly encountered by fire and emergency services personnel?

A. Hazardous materials incidents
B. Transportation (aircraft) incidents
C. Transportation (vehicle) incidents
D. Flammable and combustible liquids incidents

_____ 9. Which of the following statements about acquired structures is TRUE?

A. They must usually be purchased at fair market value from the owner.
B. They usually have no rightful owner and are considered public property.
C. They are usually in deteriorating conditions and scheduled for demolition.
D. They usually could be renovated or remodeled and sold for a profit.

_____ 10. Which of the following is NOT a responsibility of an instructor when conducting live-fire or other training evolutions?

A. Prepare water supply and flow analyses.
B. Distribute a notice to nearby residents.
C. Hold a news conference explaining the evolution.
D. Receive the appropriate permits from the local jurisdiction.

True / False

B. Write the correct letters on the blanks.

_____ 1. The communication loop must include the incident safety officer and the individual assigned to the fuel shutoff valve control.
A. True
B. False

_____ 2. Training areas do NOT need to be secured to prevent public access.
A. True
B. False

_____ 3. Driving courses should NOT include parked vehicles, overhead wires, or other obstacles that may interfere with training.
A. True
B. False

_____ 4. Mobile training facilities reduce the cost and lost time that occur when students must be transported to a central location.
A. True
B. False

_____ 5. A commercially constructed fire behavior simulator should NOT be altered without the manufacturer's permission.
A. True
B. False

_____ 6. All transportation props require the same safety and environmental requirements of other live-fire training props.
A. True
B. False

_____ 7. Because a simulated hazardous material is used in training exercises, runoff water does not need to be contained.
A. True
B. False

_____ 8. A live-fire training area must be equipped with a fuel shutoff or automatic fire-suppression system in the event of an accident.
A. True
B. False

_____ 9. Portable training props that are used to simulate gas pressure fires are referred to as "Christmas trees."

A. True
B. False

_____ 10. Acquired structures used for live-fire training will be destroyed in the process and do NOT need to meet any standards.

A. True
B. False

_____ 11. When the presence of asbestos or other hazardous materials is confirmed in an acquired structure, the building owner is required to use certified personnel to remove and dispose of the materials.

A. True
B. False

_____ 12. It may NOT be possible to use acquired structures for live-fire training because of environmental laws or the designation of a historical landmark.

A. True
B. False

_____ 13. Training exercises should never be canceled because of environmental conditions.

A. True
B. False

_____ 14. Water rescue incidents should NOT be simulated because they are inherently dangerous.

A. True
B. False

_____ 15. When using fuels in live-fire training use only fuels with known burning characteristics.

A. True
B. False

List

C. Write five characteristics that live-fire training areas must have before they can be used for any training activity on the lines provided.

1. ______________________________

____

2. ______________________________

____

3. ______________________________

____

4. ______________________________

____

5. ______________________________

____

Chapter 6 Report Writing and Record Keeping

Multiple Choice

A. Write the correct letters on the blanks.

_____ 1. Which of the following parts of a report provides a brief overview of the report in a single paragraph and includes the purpose of the report?

A. Heading
B. Body
C. Introduction
D. Conclusion/summary

_____ 2. Which of the following parts of a report contains all information relating to the report including statistics and problems that were uncovered?

A. Heading
B. Body
C. Introduction
D. Conclusion/summary

_____ 3. Which of the following parts of a report restates the results of the project briefly and recommends changes or other actions?

A. Heading
B. Body
C. Introduction
D. Conclusion/summary

_____ 4. What can be done to highlight complicated information in a report?

A. Use smaller font.
B. Place it in an appendix.
C. Present it in a table or graph.
D. Place it in a separate publication.

_____ 5. Which of the following should be avoided in report writing?

A. Using active voice
B. Giving recommendations
C. Using statistics or figures
D. Using excessively technical words

____ 6. Which of the following prohibits the release of student testing and evaluation information in the United States?

A. Second Amendment
B. Fifth Amendment
C. Title VII of the Training Act
D. Family Education and Privacy Act

____ 7. Which of the following statements about maintaining training records is TRUE?

A. Record-keeping requirements are set only by the organization.
B. Training records are considered optional and do not have requirements.
C. Training records are only required for departments with over 100 personnel.
D. Local and state or provincial requirements must be followed when maintaining training records.

____ 8. Which of the following questions would NOT be a question to answer for an audit of the record-keeping system?

A. What is the purpose of the record?
B. How secure is the information?
C. Do the records take up too much storage space?
D. What trends in training can be determined from the records?s

____ 9. Which of the following information should be retained in budget records?

A. Daily training records
B. Equipment maintenance
C. Instructor resumes
D. Purchasing records

____ 10. Which of the following records are maintained to provide data for the distribution of payrolls and benefits?

A. Incident reports
B. Purchasing records
C. Daily attendance records
D. Job performance records

____ 11. Which of the following statements about personnel records is TRUE?

A. They are generally confidential.
B. They are available to all employees.
C. They are available to the general public.
D. They are available to any government entity.

_____ 12. Which of the following is NOT a component of the actual cost for providing training?

A. Materials cost
B. Instructor certification and training costs
C. Fees and tuition paid by students
D. Prorated cost of the facility designated for training

_____ 13. Which of the following statements about computer software programs used in record-keeping is TRUE?

A. They increase the risk of a security breach.
B. They increase the time required for record keeping.
C. They are too complex for most organizations.
D. They improve the efficiency of record management.

_____ 14. Which of the following questions should be answered in a report in addition to the questions who, what, when, and where?

A. Why
B. How
C. Should
D. Would

_____ 15. Which of the following parts of a report includes the name of the recipient and the name of the sender or author?

A. Heading
B. Body
C. Introduction
D. Conclusion/summary

6

True / False

B. Write the correct letters on the blanks.

_____ 1. The majority of reports that instructors write are considered confidential and cannot be read by people outside the organization.

A. True
B. False

_____ 2. Reports should be checked for grammar and spelling and also for technical accuracy.

A. True
B. False

_____ 3. Types of activities documented by reports include training sessions, administrative activities, and emergency responses.

A. True
B. False

_____ 4. An executive summary should be lengthy and focus on personal opinions.

A. True
B. False

_____ 5. One function of reports generated by instructors is to keep an administration informed of the accomplishments, problems, and daily training activities of the divisions within an organization.

A. True
B. False

_____ 6. Inventory and fixed asset records are maintained for everything that is assigned to, consumed by, or owned by the organization.

A. True
B. False

_____ 7. Preventive maintenance is performed to repair damage to equipment because of an accident, overuse, operator error, or even abuse.

A. True
B. False

_____ 8. An executive summary is a brief review of the key points in a report, a technical paper, specifications, or an analysis.

A. True
B. False

_____ 9. General categories of records kept in the fire and emergency services include training, inventory, and maintenance.

A. True
B. False

_____ 10. Information stored in a record-keeping system is historical and cannot be used to justify budget requests or program development.

A. True
B. False

_____ 11. An executive summary should include a statement of the problem and recommended solution.

A. True
B. False

_____ 12. Types of training records include special training provided by a source outside the organization and individual training provided by an organization.

A. True
B. False

_____ 13. If accurate training records are kept, the amount of training necessary is decreased.

A. True
B. False

_____ 14. Training records provide information for planning and scheduling future training programs.

A. True
B. False

_____ 15. Information to be gathered for a training records system includes student attendance rosters, topics taught at each session, and course evaluations provided by participants.

A. True
B. False

_____ 16. The record-keeping development process includes defining the requirements, planning the system, implementing the system, and completing the project.

A. True
B. False

_____ 17. The first step in determining the record-keeping requirements of an organization is to list outcomes by listing in priority the required objectives and the desired objectives.

A. True
B. False

_____ 18. Steps in the system planning phase include establishing the final project objectives and determining the resources necessary for the project.

A. True
B. False

_____ 19. When implementing a new record-keeping system, files that already exist are not entered into the system.

A. True
B. False

_____ 20. Annual checks of a record-keeping system are not necessary if no complaints have been made.

A. True
B. False

List

C. **Write the four parts of a report on the lines provided.**

1. ____________________

2. ____________________

3. ____________________

4. ____________________

Terms

D. **Describe the following terms on the lines provided.**

1. Preventive maintenance ______________________________

____

____

2. Corrective maintenance ______________________________

____

____

3. Data ______________________________

____

____

4. Analysis ______________________________

____

____

Chapter 7 Principles of Learning

Multiple Choice

A. Write the correct letters on the blanks.

_____ 1. Which of the following terms refers to an internal state or condition that activates and directs behavior toward a goal?
A. Emotion
B. Motivation
C. Readiness
D. Learning curve

_____ 2. Which of the following terms refers to a survey of the types of services required or desired by the community or service area?
A. Public opinion
B. Needs assessment
C. Task/job analysis
D. Community analysis

_____ 3. Which of the following occurs through the various experiences that people have in life?
A. Teaching
B. Training
C. Formal learning
D. Informal learning

_____ 4. Which of the following terms refers to specific statements that describe desired learning results (competencies)?
A. Goals
B. Learning outcomes
C. Learning objectives
D. Lesson plans

_____ 5. In which of the following learning domains do students gain understanding about a concept or topic?
A. Affective domain
B. Cognitive domain
C. Primary domain
D. Psychomotor domain

_____ 6. Which of the following levels of learning in the cognitive domain involves dividing information into its component parts to understand the relationship between the parts and understand the whole?

A. Knowledge
B. Analysis
C. Synthesis
D. Comprehension

_____ 7. Which of the following levels of learning in the cognitive domain involves using information learned in new and specific situations?

A. Knowledge
B. Synthesis
C. Application
D. Evaluation

_____ 8. Which of the following levels of learning in the cognitive domain involves remembering, recalling, and recognizing previously learned facts and theories?

A. Knowledge
B. Analysis
C. Synthesis
D. Comprehension

_____ 9. Which of the following is the first step in learning in the psychomotor domain?

A. Imitation
B. Observation
C. Performance
D. Perfection

_____ 10. Which of the following is the final step in learning in the psychomotor domain?

A. Imitation
B. Adaptation
C. Performance
D. Perfection

_____ 11. Which of the following learning methods involves using a step-by-step, orderly thinking process that has both a beginning and an end?

A. Global or holistic
B. Sequential or linear
C. Abstract or symbolic
D. Concrete or real objects or items

_____ 12. Which of the following learning methods involves seeing the whole picture and forming relationships between concepts, events, or things?

A. Global or holistic
B. Sequential or linear
C. Abstract or symbolic
D. Concrete or real objects or items

_____ 13. Which of the following learning methods involves recognizing common qualities in similar but different experiences?

A. Global or holistic
B. Sequential or linear
C. Abstract or symbolic
D. Concrete or real objects or items

_____ 14. Which of the following laws of learning stresses the idea that the more an act is practiced, the faster and surer the learning becomes?

A. Readiness
B. Exercise
C. Recency
D. Effect

_____ 15. Which of the following laws of learning states the principle that the first of a series of learned acts would be remembered better than others?

A. Intensity
B. Exercise
C. Recency
D. Primacy

_____ 16. Which of the following laws of learning states that if a stimulus (experience) is vivid and real, it will more likely change or have an effect on the behavior (learning)?

A. Intensity
B. Exercise
C. Recency
D. Primacy

_____ 17. Which of the following is NOT a positive action instructors can take to motivate students?

A. Demonstrate enthusiasm
B. Provide relevancy
C. Stimulate motivation
D. Expect only average performance

_____ 18. Which of the following is NOT a characteristic of the mastery approach to teaching?
A. Competency-based
B. Instructor-supported
C. Delayed, general feedback
D. Modules and multimedia

_____ 19. Which of the following is NOT a characteristic of the traditional approach to teaching?
A. Content-based
B. Independent learning
C. Textbook/workbook
D. Delayed, general feedback

_____ 20. Which of the following is a disadvantage of the mastery approach to teaching?
A. Instructors must perform task analyses.
B. Students are prepared to advance to more complex knowledge.
C. The responsibility for learning is focused on the student.
D. Additional time must be available to ensure that all students master the subject.

True / False

B. Write the correct letters on the blanks.

_____ 1. In the comprehension level of learning in the cognitive domain students put parts together to form a new whole and they integrate parts to invent new procedures.
A. True
B. False

_____ 2. One assumption of the theory of andragogy is that adults are ready to learn whatever they need to know or do in order to meet job requirements or social roles.
A. True
B. False

_____ 3. The least understood domain of learning is the psychomotor domain.
A. True
B. False

_____ 4. Pedagogy is the principle of learning most often associated with adults.
A. True
B. False

_____ 5. The three basic categories of learning styles are audio, visual, and kinesthetic.

A. True
B. False

_____ 6. The three domains of learning are the cognitive domain, the psychomotor domain, and the affective domain.

A. True
B. False

_____ 7. The cone of learning model states that individuals retain approximately 50 percent of what they say while doing what they are talking about.

A. True
B. False

_____ 8. Affective learning involves how individuals deal with issues emotionally and includes traits such as individual awareness and attitudes.

A. True
B. False

_____ 9. One method of motivating students is to show that classroom knowledge and skills can be applied to real-life situations.

A. True
B. False

_____ 10. Long-term memory is not considered permanent storage because it may eventually be forgotten.

A. True
B. False

_____ 11. The mastery approach to teaching requires that the student successfully master the learning objectives or outcomes of the lesson or course.

A. True
B. False

_____ 12. To increase the relationship between motivation and learning instructors should gain interest and ensure success by using a variety of teaching styles that match learning styles, abilities, and needs.

A. True
B. False

_____ 13. External motivations such as rewards, recognition, and certificates do little to increase the adult student's motivation to learn.

A. True
B. False

_____ 14. Short-term memory is the memory component that holds information for about 20 seconds and is limited to about seven items or chunks of information.
A. True
B. False

_____ 15. One method of motivating students is to avoid having students work together in peer groups.
A. True
B. False

_____ 16. Instructors should relate knowledge that the students possess in their long-term memory to new information and concepts.
A. True
B. False

_____ 17. Learning plateaus are NOT normal and affect only a small percentage of students.
A. True
B. False

_____ 18. The mastery approach to teaching is NOT well suited for the fire and emergency services.
A. True
B. False

_____ 19. In the mastery approach to teaching students cannot proceed to new material until the basic requisite material is mastered.
A. True
B. False

_____ 20. The mastery approach to teaching uses criterion-referenced teaching, learning, and assessments and focuses attention on learning objectives.
A. True
B. False

_____ 21. A disadvantage of the mastery approach to teaching is that instructors must state the learning objectives before designating or designing student activities and projects.
A. True
B. False

_____ 22. An advantage of the mastery approach to teaching is that knowledge that the student possessed before the course can be used to gain mastery more quickly.
A. True
B. False

_____ 23. Criterion-referenced assessments measure the accomplishments of one student against that of another.
A. True
B. False

_____ 24. Norm-referenced assessments are the most commonly used assessments in the fire and emergency services.
A. True
B. False

_____ 25. A criterion-referenced assessment measures student performance by comparing it to the standard or criterion stated in the course objectives.
A. True
B. False

List

C. Write the levels of learning in the cognitive domain in order from simple to complex on the lines provided.

1. ____________________
2. ____________________
3. ____________________
4. ____________________
5. ____________________
6. ____________________

D. **Write the levels of learning in the psychomotor domain in order from simple to complex on the lines provided.**

1. ______________________________

2. ______________________________

3. ______________________________

4. ______________________________

5. ______________________________

Describe

E. **Describe the three domains of learning in relation to the students on the lines provided.**

1. ______________________________

____

____

2. ______________________________

____

____

3. ______________________________

____

____

Chapter 8 Student Attributes and Behaviors

Multiple Choice

A. Write the correct letters on the blanks.

_____ 1. Which of the following age categories have a tendency to bring personal concerns into the classroom?

A. Gen Xer
B. Dot com
C. Baby boomer
D. Traditionalist

_____ 2. Which of the following age categories tend to be fiscally conservative and place a high value on institutions such as universities, corporations, and religion?

A. Gen Xer
B. Dot com
C. Baby boomer
D. Traditionalist

_____ 3. Which of the following age categories composes the majority of fire and emergency services organizations?

A. Gen Xer
B. Dot com
C. Baby boomer
D. Traditionalist

_____ 4. Which of the following would NOT be an indicator of a learning disability?

A. Problems with concentration
B. Problems with auditory and visual perception
C. Emotional problems, such as the fear of rejection
D. Problems in reading, such as word recognition and comprehension

_____ 5. Which of the following terms refers to a wide variety of disorders that may be neurological in origin and affect the individual's ability to understand, think, or use the spoken or written word?

A. Literacy disorders
B. Learning disabilities
C. Inherited disabilities
D. Educational disorders

_____ 6. Which of the following is the process of giving motivational correction, positive reinforcement, and constructive feedback to students?

A. Coaching
B. Counseling
C. Mentoring
D. Positive teaching

_____ 7. Which of the following places a new student under the guidance of a more experienced professional or another student who acts as tutor, guide, and motivator?

A. Coaching
B. Counseling
C. Mentoring
D. Positive teaching

_____ 8. Which of the following would NOT be an example of student-originated disruptive behavior?

A. Arriving late
B. Seeking attention
C. Talking with others off the subject
D. Having an opinion different from the instructor

_____ 9. Which of the following is a student reaction to the instructor action of intimidation?

A. Gets fidgety
B. Feels insecure
C. Shows no interest
D. Feels unstimulated or bored

_____ 10. Which of the following is a student reaction to the instructor action of running overtime?

A. Gets fidgety
B. Feels insecure
C. Shows no interest
D. Feels unstimulated or bored

_____ 11. Which of the following age categories composes the leadership and upper ranks of fire and emergency services organizations?

A. Gen Xer
B. Dot com
C. Baby boomer
D. Traditionalist

_____ 12. Which of the following categories of students will be helped by materials that have the following characteristics: short sentences, double-spaced lines, and large type?

A. Gifted students
B. Nondisruptive, nonparticipating students
C. Disruptive, nonparticipating students
D. Individuals with low literacy levels

_____ 13. Which of the following would be the first step in dealing with a student who is talkative and aggressive?

A. Make a private appeal.
B. Call the student's counselor.
C. Expel the student from the course.
D. Make a public appeal in the classroom.

_____ 14. Signs such as glazed looks, gazing around the room, and thumbing through unrelated materials may indicate which type of student?

A. Disruptive students
B. Quiet or bored students
C. Shy or timid students
D. Talkative and aggressive students

_____ 15. Which of the following is the student reaction to the instructor action of attempting to overcontrol?

A. Rebels
B. Gets fidgety
C. Feels insecure
D. Feels unstimulated or bored

True / False

B. Write the correct letters on the blanks.

_____ 1. Instructors should not try to establish a relationship between past experiences and new information since all students' past experiences are different.

A. True
B. False

_____ 2. Adult students may be distracted from the learning process because of multiple responsibilities and obligations.

A. True
B. False

_____ 3. Educational background influences individual attitudes, confidence, and ability to handle new learning experiences.
A. True
B. False

_____ 4. Education level refers to reading and comprehension ability.
A. True
B. False

_____ 5. Gifted adult students will usually accomplish more than is expected of average students, and they may study and learn very well without much supervision.
A. True
B. False

_____ 6. Instructors should immediately start calling on shy or timid students for discussion or response to get them involved in the class.
A. True
B. False

_____ 7. When problems persist with disruptive students, the instructor should always follow the organization's discipline policies and procedures.
A. True
B. False

_____ 8. The majority of students in most fire and emergency services courses are nondisruptive, participating, and successful.
A. True
B. False

_____ 9. In an adult learning environment, the instructor does NOT need to perform behavior management because peer pressure will control the classroom.
A. True
B. False

_____ 10. Items to be covered when reviewing policies include safety rules, facility layout, class participation, and methods of evaluation.
A. True
B. False

_____ 11. It is acceptable for instructors to assume the role of therapist when a student appears to have a psychological/emotional problem.

A. True
B. False

_____ 12. When an instructor must interrupt class to manage disruptive behavior, the event should be documented for reference and the record.

A. True
B. False

_____ 13. Peer assistance is a process that involves having students assist other students in the learning process.

A. True
B. False

_____ 14. Counseling sessions between students and instructors must be done in private.

A. True
B. False

_____ 15. If a student's disruptive behavior does not stop after all other options are exhausted, the instructor may expel the student from the class.

A. True
B. False

List

C. Write examples of student-originated disruptive behavior on the lines provided.

1. ______________________________
2. ______________________________
3. ______________________________
4. ______________________________
5. ______________________________
6. ______________________________
7. ______________________________
8. ______________________________
9. ______________________________

Terms

D. Describe the following terms on the lines provided.

1. Learning disabilities ____________________

2. Coaching ____________________

3. Mentoring ____________________

4. L-E-A-S-T method of progressive discipline ____________________

Chapter 9 Preparation for Instruction

Multiple Choice

A. Write the correct letters on the blanks.

_____ 1. Which of the following steps in the four-step method of instruction has the purpose of determining whether students achieved the lesson objectives?
A. Preparation
B. Presentation
C. Application
D. Evaluation

_____ 2. Which of the following components of a lesson plan provides a description of the minimum acceptable behaviors that a student must display by the end of an instructional period?
A. Course goals
B. Assignments
C. Prerequisites
D. Learning objectives

_____ 3. Which of the following components of a lesson plan provides a list of information, skills, or previous requirements that students must have completed or mastered before entering the course or starting the lesson?
A. Assignments
B. References
C. Prerequisites
D. Lesson outline

_____ 4. Which of the following components of a lesson plan provides a summary of the information to be taught?
A. Evaluations
B. References
C. Prerequisites
D. Lesson outline

_____ 5. Which of the following steps in the four-step method of instruction provides opportunities for learning through activities, exercises, discussions, and skill practices?
A. Preparation
B. Presentation
C. Application
D. Evaluation

_____ 6. Which of the following steps in the four-step method of instruction is intended to motivate students to learn?

A. Preparation
B. Presentation
C. Application
D. Evaluation

_____ 7. In which of the following steps in the four-step method of instruction does the instructor present the information using an orderly, sequential outline?

A. Preparation
B. Presentation
C. Application
D. Evaluation

_____ 8. Which of the following seating arrangements arranges students in fixed seating that permanently faces the stage or lectern and permits only the interaction between students and instructor?

A. Fan
B. Chevron
C. Auditorium or theater
D. Horseshoe or U-shape

_____ 9. Which of the following seating arrangements permits students to easily see and hear the instructor and works effectively in small groups?

A. Fan
B. Chevron
C. Auditorium or theater
D. Horseshoe or U-shape

_____ 10. Which of the following seating arrangements provides space for meal functions, discussion groups, or small group meetings?

A. Conference
B. Hollow square
C. Round tables
D. Circled chairs

_____ 11. Which of the following seating arrangements is best used for small-to-medium-sized groups where discussion is the primary method of teaching and is NOT useful when students are expected to take notes?

A. Conference
B. Hollow square
C. Round tables
D. Circled chairs

_____ 12. Which of the following steps in the four-step method of instruction has the purpose of reinforcing the student's learning and is typically related to performing the operations of a task?

A. Preparation
B. Presentation
C. Application
D. Evaluation

_____ 13. Which of the following two steps in the four-step method of instruction are often combined?

A. Preparation and presentation
B. Presentation and application
C. Application and evaluation
D. Evaluation and preparation

_____ 14. Which of the following components of a lesson plan consists of reading, practice, research, or other outside-of-class requirements for students?

A. Assignments
B. References
C. Prerequisites
D. Resources/material needed

_____ 15. Which of the following components of a lesson plan consists of a short descriptive title of the information covered?

A. References
B. Job or topic
C. Lesson outline
D. Level of instruction

True / False

B. Write the correct letters on the blanks.

_____ 1. An instructor who is unqualified to teach a topic should teach the topic only if using a qualified instructor's lesson plans.

A. True
B. False

_____ 2. In the four-step method of instruction most learning takes place in the application step.

A. True
B. False

_____ 3. The instructor is responsible for ensuring that all materials and equipment needed are determined and arranged for before the beginning of class.
A. True
B. False

_____ 4. One of the primary roles of the Level I instructor is to develop new lesson plans.
A. True
B. False

_____ 5. A lesson plan is an instructional tool that establishes the steps that an instructor will take to complete the various objectives and ultimate goal of the course.
A. True
B. False

_____ 6. The amount of enthusiasm projected by an instructor directly affects the level of enthusiasm and interest exhibited by students.
A. True
B. False

_____ 7. Logistical needs, such as restocking kits with supplies, may consume class time.
A. True
B. False

_____ 8. One method of ensuring course continuity is to prepare students for different instructors by giving the students some background on the substitute's experiences, knowledge, and teaching methods.
A. True
B. False

_____ 9. Courses that are scheduled during the times of year when inclement weather is possible must have some flexibility built into them.
A. True
B. False

_____ 10. The equipment used in learning sessions can differ from the equipment used in testing and on the job.
A. True
B. False

_____ 11. Instructors must adapt lessons with various activities and formats so that every student can gain the appropriate knowledge and skill and meet the lesson objectives.

A. True
B. False

_____ 12. Following safety procedures in training exercises is of lower importance than following safety procedures during emergency incidents.

A. True
B. False

_____ 13. Frequent breaks during instruction are NOT necessary for adult students since they are able to sit for long periods of time.

A. True
B. False

_____ 14. In the indoor learning environment, the instructor should have control over seating, lighting, temperature, noise level, and audiovisual equipment.

A. True
B. False

_____ 15. The primary source of light for an indoor classroom should be incandescent lights.

A. True
B. False

_____ 16. Instructors must always preview audiovisual aids before presenting the material.

A. True
B. False

_____ 17. It is NOT the instructor's responsibility to eliminate safety hazards, such as tripping hazards and electrical cords, in the indoor classroom.

A. True
B. False

_____ 18. Outdoor training should NOT be conducted along public streets or in parking lots that may be affected by vehicle traffic.

A. True
B. False

_____ 19. At the beginning of any outdoor training session, instructors must provide an overview of the training scenario, indicating expected outcomes, safety issues, and unit or company or individual assignments.

A. True
B. False

_____ 20. Even realistic training scenarios should NOT be as noisy as actual emergency incidents.

A. True
B. False

List

C. Write the steps in the four-step method of instruction on the lines provided.

1. ______________________________
2. ______________________________
3. ______________________________
4. ______________________________

D. Write the considerations an instructor should address during the inspection and planning processes for outside instruction on the lines provided.

1. ______________________________
2. ______________________________
3. ______________________________
4. ______________________________
5. ______________________________
6. ______________________________
7. ______________________________
8. ______________________________
9. ______________________________

Describe

E. Describe the following basic components of a lesson outline on the lines provided.

1. Job or topic ______________________________
____
2. Time frame ______________________________
____
3. Level of instruction ______________________________
____
____
4. Learning objectives ______________________________
____
____
5. Resources/materials needed ______________________________
____
____
6. Prerequisites ______________________________
____
____
7. References ______________________________
____
____

8. Lesson summary ______________________________

____

____

9. Assignments ______________________________

____

____

10. Lesson outline ______________________________

____

____

11. Evaluations ______________________________

____

____

Chapter 10 Instructional Delivery

Multiple Choice

A. Write the correct letters on the blanks.

_____ 1. Which of the following is a disadvantage of the lecture format?
A. Time-efficient
B. Cost-effective
C. Familiar to students
D. Limited student/instructor interaction

_____ 2. Which of the following is NOT a guideline for using visual aids in an illustrated lecture?
A. Illustrate a single lesson objective in each visual.
B. Make the visual large enough to be easily seen.
C. Have visuals displayed during the entire lesson.
D. Display steps in sequence individually when illustrating the steps in an operation.

_____ 3. Which of the following is a disadvantage of the demonstration presentation format?
A. Participants can receive feedback immediately.
B. Instructors can readily observe a change in behavior.
C. Students have a high level of interest when participating.
D. Instructors must plan for extensive preparation and cleanup times.

_____ 4. Which of the following is a reason that technology-based training is becoming increasingly popular?
A. Increase in funding for training budgets
B. Decrease in the number of nontraditional students
C. Improved sophistication in computer-based simulations
D. Decrease in demand for specialized courses with limited enrollment

_____ 5. Which of the following is an advantage of technology-based training?
A. Requires students to have good writing skills
B. Requires a high level of student self-discipline and self-motivation
C. Course preparation and participation times are greater than classroom time
D. Changes and updates to course material can be made in real time

_____ 6. Which of the following is a guideline for giving an effective presentation?
A. Avoid eye contact with students while speaking.
B. Begin quickly with new information to keep up interest.
C. Avoid pausing while giving a presentation.
D. Be yourself and use your own unique style, experiences, and abilities as a person.

_____ 7. Which of the following component of an oral presentation is used to get the attention of students?
A. Introduction
B. Body
C. Summary
D. Transitions

_____ 8. Which of the following sequencing methods begins with an overview of the entire topic or a demonstration of the complete skill in real time?
A. Whole-to-part
B. Simple-to-complex
C. Part-to-whole
D. Known-to-unknown

_____ 9. Which of the following sequencing methods begins by teaching the basic knowledge or skills and then introducing more difficult or complex knowledge as the lesson progresses?
A. Whole-to-part
B. Simple-to-complex
C. Part-to-whole
D. Known-to-unknown

_____ 10. Which of the following sequencing methods works best when teaching a process that is composed of individual steps?
A. Chronological
B. Simple-to-complex
C. Step-by-step
D. Known-to-unknown

_____ 11. Which of the following sequencing methods begins by describing a part, such as the impeller of a pump, and then shows how the part works with the larger unit, in this case the pump?
A. Whole-to-part
B. Simple-to-complex
C. Part-to-whole
D. Known-to-unknown

_____ 12. Which of the following is NOT an advantage of active learning?

A. Fosters improved student understanding
B. Improves cooperation within a group
C. Improves student communication skills
D. Places the responsibility for learning in the hands of the instructor

_____ 13. Which of the following is NOT a strategy for stimulating student interest?

A. Stimulate emotions.
B. Relate learning to student interests.
C. Use questions to stimulate interest.
D. Provide extra-credit for participation.

_____ 14. Which of the following types of questions has a limited number of possible answers?

A. Open
B. Closed
C. Relay
D. Rhetorical

_____ 15. Which of the following types of questions is used to stimulate thinking or motivate participants and does not necessarily have an oral response?

A. Open
B. Closed
C. Relay
D. Rhetorical

True / False

B. Write the correct letters on the blanks.

_____ 1. An advantage of the lecture format is that there are limited senses involved in receiving the information.

A. True
B. False

_____ 2. It is NOT necessary for students to have a basic knowledge of the subject before the discussion begins.

A. True
B. False

_____ 3. Technology-based training is electronic learning that uses methods such as Internet web-based instruction, interactive television (ITV), and other forms of computer-based electronically transferred knowledge.
A. True
B. False

_____ 4. Methods to overcome the disadvantages of the lecture format include adding discussions, illustrations, demonstrations, and structured exercises into the lecture.
A. True
B. False

_____ 5. The discussion presentation format works well with large groups of approximately 40-60 students.
A. True
B. False

_____ 6. In a guided discussion the instructor presents a topic to a group and the members of the group discuss ideas in an orderly exchange controlled by the instructor.
A. True
B. False

_____ 7. A guideline for preparing for a demonstration is to acquire all equipment and accessories, ensure that they work, and arrange them for use.
A. True
B. False

_____ 8. When demonstrating a skill do NOT allow students the opportunity to ask questions.
A. True
B. False

_____ 9. When demonstrating a skill begin by linking new information with the students' current knowledge.
A. True
B. False

_____ 10. The multiple-instructor presentation format requires fewer hours per instructor to accomplish the same course requirements and requires more advanced planning than individual teaching.
A. True
B. False

_____ 11. A disadvantage of the multiple-instructor presentation format is that it does NOT work well when the topic is broad.
A. True
B. False

_____ 12. An advantage of the multiple-instructor presentation format is that it provides students with an exposure to a wide variety of teaching methods and skills training.
A. True
B. False

_____ 13. Blended electronic-learning combines online learning courses that students complete independently with classroom-delivered instruction and hands-on, performance-based skills instruction.
A. True
B. False

_____ 14. Interactive television is used to link multiple classroom sites together and permits one instructor to reach more students.
A. True
B. False

_____ 15. Security is NOT a major issue with technology-based training because the students are adults.
A. True
B. False

_____ 16. In self-directed learning an instructor is directly involved in the delivery of the training.
A. True
B. False

_____ 17. All types of training programs, particularly basic-level skill programs, are suited to self-directed learning.
A. True
B. False

_____ 18. Self-directed learning places the responsibility for achieving the course objectives solely on the student.
A. True
B. False

_____ 19. Characteristics of effective speakers include being audience-centered and using appropriate vocal characteristics.
A. True
B. False

_____ 20. The transmission of information for the affective domain occurs through teaching and through personal actions.
A. True
B. False

_____ 21. One strategy to promote active learning is to intersperse small lecture segments with discussion groups or skills practice times.
A. True
B. False

_____ 22. Repetition takes up class time and does NOT help to reinforce learning.
A. True
B. False

_____ 23. Guidelines for asking effective questions include allowing a wait time and asking questions at a variety of levels and types.
A. True
B. False

_____ 24. Guidelines for asking effective questions include asking several questions at a time and using questions to embarrass students who are not paying attention.
A. True
B. False

_____ 25. When an instructor does not know the answer to a question, it is appropriate to bluff students so that class can continue.
A. Truc
B. False

Terms

C. Describe the following terms on the lines provided.

1. Instructor-led training ______________________________

____

____

2. Technology-based training ______________________________

____

____

3. Blended electronic learning ______________________________

____

____

4. Individualized instruction ______________________________

____

____

5. Active learning ______________________________

____

____

Chapter 11 Audiovisual Technology

Multiple Choice

A. Write the correct letters on the blanks.

_____ 1. Which of the following is NOT a benefit of using audiovisual aids?
- A. Increase lecture time
- B. Add interest to a lecture
- C. Help students organize ideas
- D. Enhance student understanding

_____ 2. Which of the following is a guideline for using audiovisual training aids?
- A. Use as many training aids as possible.
- B. Do NOT spend a lot of preparation time for visual aids.
- C. Rehearse with the training aids before giving a lecture.
- D. Use visual aids that are complex and present multiple ideas.

_____ 3. Which of the following is a strategy for avoiding distractions when using audiovisual training aids?
- A. Avoid eye contact with students when using a visual aid.
- B. Use multiple training aids simultaneously whenever possible.
- C. Introduce audiovisual training aids at the time they are to be viewed.
- D. Display projected visual training aids below the eye level of seated students.

_____ 4. Which of the following is NOT a purpose of transitions?
- A. Maintain consistency
- B. Provide previews
- C. Provide summaries
- D. Increase lecture time

_____ 5. Which of the following words should be avoided in verbal transitions?
- A. Finally
- B. Not only
- C. In addition to
- D. In other words

_____ 6. Which of the following is the easiest, most frequently used, and most versatile nonprojection-type equipment?
A. Models
B. Audiotapes
C. Marker board illustrations
D. Casualty simulation training aids

_____ 7. Which of the following scans and prints a reduced-size copy of material drawn or printed on the board and is a versatile variation of the dry-erase marker board?
A. Interactive board
B. Peripheral board
C. Electronic copyboard
D. Multimedia board

_____ 8. Which of the following is an excellent medium for illustrating mechanical or spatial concepts?
A. Models
B. Easel pads
C. Duplicated materials
D. Illustration or diagram displays

_____ 9. Which of the following give tremendous benefits in increasing realism for EMS training involving hands-on applications?
A. Models
B. Easel pads
C. Illustration or diagram displays
D. Casualty simulation training aids

_____ 10. Which of the following is an advantage of projected audiovisual training aids?
A. Stimulate multiple senses simultaneously
B. Large investment in audiovisual equipment
C. Costly purchase of projected training-aid presentations
D. Extensive time spent in creating presentations

_____ 11. Which of the following is NOT a guideline for video presentations?
A. Always preview the video before showing it.
B. Leave the room only briefly while a class views a video.
C. Do NOT use a video that takes more than half of the class session.
D. Emphasize key learning points/objectives before showing a video.

_____ 12. Which of the following consists of a small video camera mounted vertically over a tabletop platform?
A. Video projectors
B. Multimedia projectors
C. Portable presenters
D. Visual presenters/displays

_____ 13. Which of the following refers to a simulation that displays a field of view as if the participant is part of the simulated environment?
A. Electronic simulator
B. Smoke simulator
C. Virtual reality
D. Display board

_____ 14. Which of the following can be constructed from parts removed from scrapped or salvaged vehicles?
A. Electronic simulator
B. Smoke simulator
C. Virtual reality
D. Display board

_____ 15. Which of the following statements about videos is TRUE?
A. Video recording and editing equipment is still extremely expensive.
B. Images captured on video can only be stored on digital tapes.
C. Television has limited potential as an audiovisual training aid device.
D. Videos are commercially available on a tremendous variety of fire and emergency service subjects.

True / False

B. Write the correct letters on the blanks.

_____ 1. Class size and expected level of interaction are key factors in selecting audiovisual training aids.
A. True
B. False

_____ 2. Standardized training curricula has made it more difficult for instructors to select appropriate audiovisual training aids.
A. True
B. False

_____ 3. Nonverbal transitions may consist of a change of facial expression, a pause, a change in vocal pitch or rate of speaking, a gesture, or physically moving from one point to another within the space.

A. True
B. False

_____ 4. Instructors should have a contingency plan available in case there is a failure with the audiovisual training aid device.

A. True
B. False

_____ 5. A guideline for media transitions is to keep backgrounds complex so that they can grab attention.

A. True
B. False

_____ 6. Reasons that transitions are used include ending one topic and beginning another and moving from one teaching method into another.

A. True
B. False

_____ 7. A guideline for media transitions is to use variety in the composition of the various elements on slides by interspersing graphs, charts, photographs, and clipart to create interest.

A. True
B. False

_____ 8. Rear-screen projection systems are less expensive than front-screen projection devices.

A. True
B. False

_____ 9. An advantage of nonprojected audiovisual training aids over projected training aids is that they do NOT depend on high levels of technology or technical skill.

A. True
B. False

_____ 10. Level I instructors do NOT need to be familiar with equipment used in the completion of their duties

A. True
B. False

_____ 11. Ancillary equipment consists of those devices such as televisions, projection screens, cameras, scanners, and video-capture devices that are used in support of audiovisual training aids.

A. True
B. False

_____ 12. Electrical power cords can be left plugged in while opening audiovisual equipment.

A. True
B. False

_____ 13. Air filters in multimedia projectors do NOT need to be cleaned.

A. True
B. False

_____ 14. Instructors who are familiar with the equipment may perform minor maintenance on audiovisual training aids.

A. True
B. False

_____ 15. Cleaning guidelines for audiovisual training aid devices include using only manufacturer-recommended cleaning agents and procedures.

A. True
B. False

_____ 16. Cleaning guidelines for audiovisual training aid devices include cleaning chalk and whiteboards completely when finished using them.

A. True
B. False

_____ 17. Care guidelines for audiovisual training aids and devices include avoiding placing dust covers over equipment when not in use.

A. True
B. False

_____ 18. Care guidelines for audiovisual training aid devices include NOT leaving electronic equipment in vehicles when temperature extremes are expected.

A. True
B. False

_____ 19. Care guidelines for audiovisual training aid devices include storing mannequins properly in carrying cases or closed cabinets.
A. True
B. False

_____ 20. Cleaning guidelines for audiovisual training aid devices include using solvents for cleaning training aids.
A. True
B. False

List

C. Write the purposes of transitions on the lines provided.

1. ______________________________
2. ______________________________
3. ______________________________
4. ______________________________
5. ______________________________
6. ______________________________

Chapter 12 Structured Exercises, Demonstrations, and Practical Training Evolutions

Multiple Choice

A. Write the correct letters on the blanks.

_____ 1. Which of the following structured exercises is a description of a real or hypothetical problem that an organization or an individual has dealt with or could deal with?

A. Simulation
B. Case study
C. Brainstorming sessions
D. Field and laboratory experiences

_____ 2. Which of the following structured exercises has students portray characters and act the roles assigned to them in scenarios?

A. Simulation
B. Case study
C. Role-play
D. Field and laboratory experiences

_____ 3. Which of the following structured exercises involves a group of students given a problem or situation and then given time to determine a solution to it?

A. Simulation
B. Case study
C. Brainstorming sessions
D. Field and laboratory experiences

_____ 4. Which of the following structured exercises are particularly helpful when the aspects of the affective domain need to be taught or reinforced?

A. Simulation
B. Case study
C. Role-play
D. Field and laboratory experiences

_____ 5. Which of the following structured exercises is an activity that allows students to participate in a scenario that represents a real-life situation?

A. Simulation
B. Case study
C. Role-play
D. Brainstorming sessions

_____ 6. Which of the following structured exercises permits students to experience a situation and see the results of their decisions without the negative results that can occur at an actual emergency?

A. Simulation
B. Case study
C. Role-play
D. Brainstorming sessions

_____ 7. Which of the following actions is included in the first step in a skills demonstration?

A. Explanation of the skill
B. Students practicing the steps
C. Demonstration of the skill at normal speed
D. Demonstration of the skill slowly

_____ 8. Which of the following is NOT a factor that the instructor must consider when preparing practical training evolutions?

A. Instructor preparation
B. Safety and health considerations
C. Location for the type of training evolution
D. Dates of similar training in other jurisdictions

_____ 9. Which of the following methods used to control an evolution refers to applying the elements of the NIMS ICS to control the evolution as though it was an actual emergency situation?

A. Teaching
B. Managing
C. Monitoring
D. Supervising

_____ 10. Which of the following methods used to control an evolution refers to observing the progress of the evolution to ensure that all lesson objectives are performed and accomplished?

A. Teaching
B. Managing
C. Monitoring
D. Supervising

_____ 11. Which of the following statements about wildland fires is TRUE?

A. Wildland fires are not a threat to urban-interface areas.
B. Wildland fires always fall within the jurisdiction of forestry or park agencies.
C. Fire departments composed of structural firefighters are NOT responsible for wildland fires, even within their jurisdiction.
D. Training for wildland fires focuses on the extreme dangers that wildland fires can pose.

_____ 12. Which NFPA standard must all flammable/combustible liquid fire training evolutions conform to?

A. NFPA 1500
B. NFPA 1403
C. NFPA 1420
D. NFPA 1703

_____ 13. Which of the following statements about large-quantity Class A material fires is TRUE?

A. Only jurisdictions with large populations are at risk for these fires.
B. In general, large-quantity Class A materials fires general a small amount of heat.
C. Creating training evolutions to match Class A material fires can be done relatively easily.
D. Potential sites for large-quantity Class A materials fires include lumberyards, refuse dumps, automobile salvage yards, and apartment complexes under construction.

_____ 14. Which of the following types of training involves defensive driving skills, pump operations, aerial device operations, auxiliary power equipment operation, and preventive care and maintenance?

A. Vehicle extrication
B. Disaster management operations
C. Emergency vehicle operations
D. Emergency medical or triage operations

_____ 15. Which of the following types of training involves the use of power and hand tools to remove a victim trapped in a wrecked vehicle?

A. Vehicle extrication
B. Disaster management operations
C. Emergency vehicle operations
D. Emergency medical or triage operations

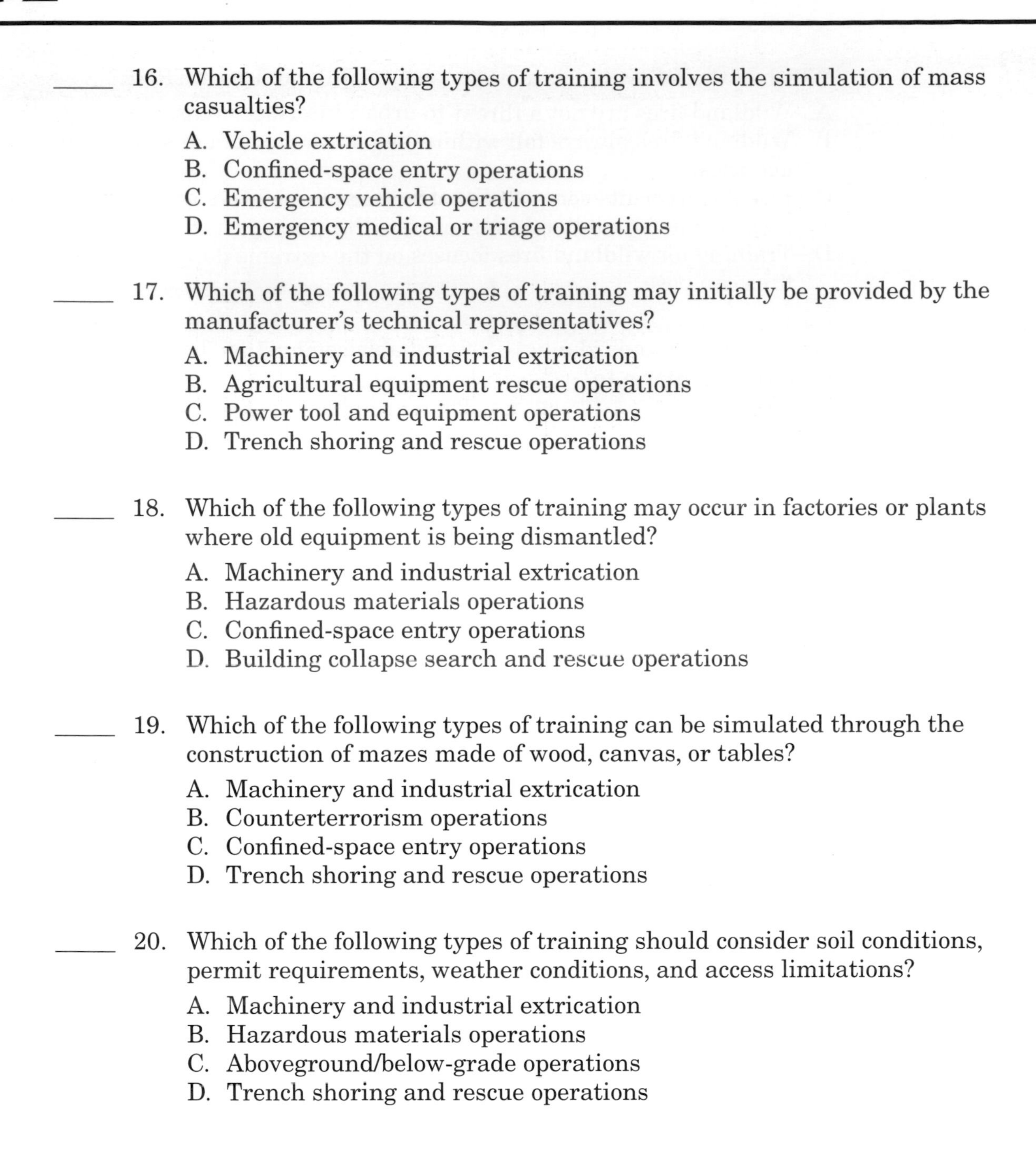

_____ 16. Which of the following types of training involves the simulation of mass casualties?

A. Vehicle extrication
B. Confined-space entry operations
C. Emergency vehicle operations
D. Emergency medical or triage operations

_____ 17. Which of the following types of training may initially be provided by the manufacturer's technical representatives?

A. Machinery and industrial extrication
B. Agricultural equipment rescue operations
C. Power tool and equipment operations
D. Trench shoring and rescue operations

_____ 18. Which of the following types of training may occur in factories or plants where old equipment is being dismantled?

A. Machinery and industrial extrication
B. Hazardous materials operations
C. Confined-space entry operations
D. Building collapse search and rescue operations

_____ 19. Which of the following types of training can be simulated through the construction of mazes made of wood, canvas, or tables?

A. Machinery and industrial extrication
B. Counterterrorism operations
C. Confined-space entry operations
D. Trench shoring and rescue operations

_____ 20. Which of the following types of training should consider soil conditions, permit requirements, weather conditions, and access limitations?

A. Machinery and industrial extrication
B. Hazardous materials operations
C. Aboveground/below-grade operations
D. Trench shoring and rescue operations

True / False

B. Write the correct letters on the blanks.

_____ 1. Experienced firefighters do NOT benefit from practical training evolutions.
A. True
B. False

_____ 2. Realistic practical training evolutions promote enthusiasm, morale, and team spirit among fire and emergency service responders.
A. True
B. False

_____ 3. Training evolutions must be held at permanent training facilities.
A. True
B. False

_____ 4. Practical training evolutions are intended to provide training in the areas of fire suppression and also technical areas such as hazardous materials.
A. True
B. False

_____ 5. When planning practical training evolutions, realism should take precedence over safety.
A. True
B. False

_____ 6. The practical training evolution must result in students meeting the learning objectives of the lesson.
A. True
B. False

_____ 7. The National Incident Management System (NIMS) ICS must be established and followed only for evolutions involving multiple agencies and jurisdictions.
A. True
B. False

_____ 8. The instructor-to-student ratio remains the same for all practical training evolutions.
A. True
B. False

_____ 9. A safety officer should be assigned at practical training evolutions to monitor all training activities.

A. True
B. False

_____ 10. Rest and rehabilitation facilities must be provided for all participants during practical training evolutions.

A. True
B. False

_____ 11. While monitoring the practical training evolution, the instructor should wait until the evolution is complete to correct any performance weaknesses or errors.

A. True
B. False

_____ 12. When more than one student is involved in a simple training evolution, the students should rotate positions so each has the opportunity to experience and practice each part of the skill.

A. True
B. False

_____ 13. In 2004, the U.S. government adopted ICS as part of NIMS, and it must be used by all federal agencies or agencies that receive federal funding.

A. True
B. False

_____ 14. All practical training evolutions that involve interior fire fighting must include the establishment of a rapid intervention team or crew (RIT/RIC).

A. True
B. False

_____ 15. Live-fire training is NOT an essential part of firefighter training because of its inherent risk.

A. True
B. False

_____ 16. All live-fire training evolutions must meet the requirements of the appropriate sections of the current NFPA 1403 standard.

A. True
B. False

_____ 17. The majority of fires involve large quantities of fuel.
A. True
B. False

_____ 18. During interior structural fire training, it is acceptable for instructors to act as victims within the burn structure.
A. True
B. False

_____ 19. During interior structural fire training it is NOT necessary for students to wear full PPE.
A. True
B. False

_____ 20. Lifelines, thermal protection, and personal flotation devices are mandatory for ice rescue training.
A. True
B. False

_____ 21. During agricultural equipment rescue operations training, either instructors or students act as victims.
A. True
B. False

_____ 22. The expanding role of fire and emergency services in protecting life and property has increased the demand for technical training evolutions involving activities other than fire suppression.
A. True
B. False

_____ 23. Training for some technical evolutions may be mandated due to government requirements.
A. True
B. False

_____ 24. Once an instructor has received training on a particular evolution, refresher courses are NOT necessary.
A. True
B. False

_____ 25. The knowledge and skills required to perform technical tasks at emergency scenes is very specialized.
A. True
B. False

List

C. Write the factors to consider when planning practical evolutions on the lines provided.

1. ______________________________
____
2. ______________________________
____
3. ______________________________
____
4. ______________________________
____
5. ______________________________
____
6. ______________________________
____
7. ______________________________
____
8. ______________________________
____
9. ______________________________
____
10. ______________________________
____
11. ______________________________
____

Describe

D. Describe the following structured exercises on the lines provided.

1. Case study ____
____
____

2. Role-plays ____
____
____

3. Brainstorming sessions ____
____
____

4. Simulations ____
____
____

5. Field and laboratory experiences ____
____
____

Chapter 13 Student Progress Evaluation and Testing

Multiple Choice

A. Write the correct letters on the blanks.

_____ 1. Which of the following terms refers to the extent to which a test measures what it was designed to measure?
A. Validity
B. Reliability
C. Evaluation
D. Measurement

_____ 2. Which of the following refers to a condition that ensures that the test is dependable by providing the same results every time it is administered?
A. Validity
B. Reliability
C. Evaluation
D. Measurement

_____ 3. Which of the following tests is given at the beginning of instruction to establish the student's current level of knowledge in order to measure readiness or determine placement?
A. Formative
B. Prescriptive
C. Summative
D. Evaluative

_____ 4. Which of the following tests are often viewed as quizzes that are given throughout the course or unit of instruction?
A. Formative
B. Prescriptive
C. Summative
D. Evaluative

_____ 5. Which of the following tests measure student achievement in an entire area covered over a long period of time?
A. Formative
B. Prescriptive
C. Summative
D. Evaluative

_____ 6. Which of the following written test questions gives students a fifty-fifty chance of guessing the correct answer?

A. Essay
B. Matching
C. True/false
D. Fill in the blank/completion

_____ 7. Which of the following written test questions consists of a series of words, dates, events, or items listed in one column and a second column that contains the definition or related information necessary to describe each item?

A. Essay
B. Matching
C. True/false
D. Fill in the blank/completion

_____ 8. Which of the following written test questions may ask students to list items, describe a process, or explain a procedure?

A. Completion
B. Matching
C. True/false
D. Short answer

_____ 9. Which of the following is NOT a guideline for administering written tests?

A. Maintain security of tests at all times.
B. Eliminate loud talking or noises outside the room.
C. Report to the assigned testing location when the test is scheduled to begin.
D. Rearrange classroom seating when necessary so that it is conducive for taking written tests.

_____ 10. Which of the following is NOT a guideline for administering performance tests?

A. Explain the purpose of each test.
B. Keep all test scores confidential.
C. Vary the manner in which you give the test to each student.
D. Provide rehabilitation facilities for students, instructors, and test evaluators.

True / False

B. Write the correct letters on the blanks.

_____ 1. Norm-referenced tests compare performance against appropriate minimum standards.
A. True
B. False

_____ 2. Oral tests are the most common test used in the fire and emergency services.
A. True
B. False

_____ 3. Evaluation refers to a process used to assess a student's achievements and/or the effectiveness of learning experiences.
A. True
B. False

_____ 4. The purpose of student evaluations and testing is to determine how well students have learned and retained the material taught to them.
A. True
B. False

_____ 5. Criterion-referenced tests compare performance against appropriate minimum standards.
A. True
B. False

_____ 6. Oral tests can be used as the sole means of evaluating students for terminal performance or officer candidates for promotion.
A. True
B. False

_____ 7. Oral tests are usually given one-on-one between a student and an instructor.
A. True
B. False

_____ 8. Written tests are always objective.
A. True
B. False

_____ 9. During performance tests, students are tested on their potential.
A. True
B. False

_____ 10. When administering written tests to adult students, it is NOT necessary to monitor students for incidents of cheating.
A. True
B. False

_____ 11. Performance tests must be based on standard criteria and performance objectives.
A. True
B. False

_____ 12. Scoring performance tests is a very objective process.
A. True
B. False

_____ 13. When scoring written tests, do NOT add comments to essay or short-answer questions.
A. True
B. False

_____ 14. When scoring performance tests, develop checklists for each skill that is tested and then use them for scoring students.
A. True
B. False

_____ 15. Student results of oral, written, and performance tests must be kept confidential.
A. True
B. False

_____ 16. Privacy of student records shall be maintained based on department/agency policies and applicable laws.
A. True
B. False

_____ 17. Students who fail to achieve the minimum required grade should be temporarily certified as having completed the training.
A. True
B. False

_____ 18. To maintain test security, instructors should remain in the room during the test and be aware of how students are taking the test.
A. True
B. False

_____ 19. Positive feedback should be timely, encouraging, specific, and easily understood.
A. True
B. False

_____ 20. Feedback should be one-sided with instructors providing the criticism and the solution.
A. True
B. False

_____ 21. Positive feedback stresses the strengths of a student's attempt at completing an activity or performance evaluation.
A. True
B. False

_____ 22. Feedback can have a negative result when it is not provided correctly.
A. True
B. False

_____ 23. In order to maintain test security, it is acceptable to rely on questions published either in a textbook or study guide.
A. True
B. False

_____ 24. In order to maintain test security, once test questions have been written do NOT attempt to revise them.
A. True
B. False

_____ 25. The increasing popularity and availability of computers provides another method for administering written tests.
A. True
B. False

Terms

C. Describe the following terms on the lines provided.

1. Criterion-referenced tests ______________________________

____

____

2. Norm-referenced tests ______________________________

____

____

3. Prescriptive tests ______________________________

____

____

4. Formative tests ______________________________

____

____

5. Summative tests ______________________________

____

____

Chapter 14 Lesson Plan Development

Multiple Choice

A. Write the correct letters on the blanks.

_____ 1. Which of the following parts of a lesson plan contains components such as the job or topic, time frame, and level of instruction?

A. Lesson outline
B. Evaluation
C. Class schedule
D. Preparation information

_____ 2. Which of the following terms is a description of the minimum acceptable behavior that a student must display by the end of an instructional period?

A. Prerequisites
B. Learning goal
C. Learning objective
D. Level of instruction

_____ 3. Which of the following parts of a lesson plan contains information that the instructor plans to use to determine whether students have met lesson objectives?

A. Lesson outline
B. Evaluation
C. Class schedule
D. Preparation information

_____ 4. Which of the following parts of a lesson plan contains the information and skills that are to be taught using the four-step method format?

A. Lesson outline
B. Evaluation
C. Class schedule
D. Preparation information

_____ 5. Which of the following statements is NOT a purpose or benefit of lesson plans?

A. Provide uniformity by standardizing the instruction
B. Give a clear path for both instructors and students to follow
C. Provide documentation for the training division
D. Enable unqualified instructors to teach a class, if needed

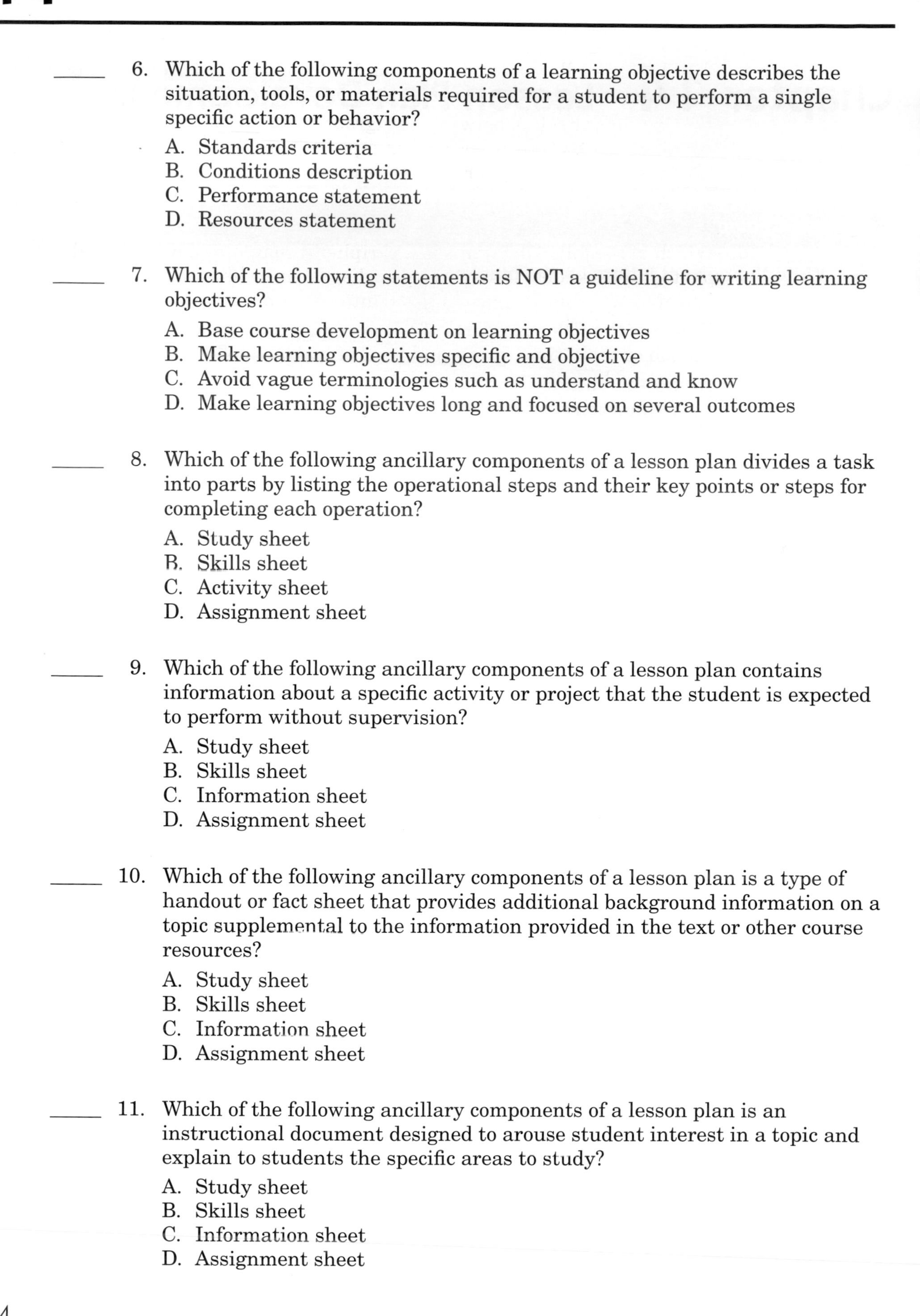

_____ 6. Which of the following components of a learning objective describes the situation, tools, or materials required for a student to perform a single specific action or behavior?
A. Standards criteria
B. Conditions description
C. Performance statement
D. Resources statement

_____ 7. Which of the following statements is NOT a guideline for writing learning objectives?
A. Base course development on learning objectives
B. Make learning objectives specific and objective
C. Avoid vague terminologies such as understand and know
D. Make learning objectives long and focused on several outcomes

_____ 8. Which of the following ancillary components of a lesson plan divides a task into parts by listing the operational steps and their key points or steps for completing each operation?
A. Study sheet
B. Skills sheet
C. Activity sheet
D. Assignment sheet

_____ 9. Which of the following ancillary components of a lesson plan contains information about a specific activity or project that the student is expected to perform without supervision?
A. Study sheet
B. Skills sheet
C. Information sheet
D. Assignment sheet

_____ 10. Which of the following ancillary components of a lesson plan is a type of handout or fact sheet that provides additional background information on a topic supplemental to the information provided in the text or other course resources?
A. Study sheet
B. Skills sheet
C. Information sheet
D. Assignment sheet

_____ 11. Which of the following ancillary components of a lesson plan is an instructional document designed to arouse student interest in a topic and explain to students the specific areas to study?
A. Study sheet
B. Skills sheet
C. Information sheet
D. Assignment sheet

_____ 12. Which of the following statements is NOT a guideline for developing effective visual aids?

A. Use similar colors between backgrounds and text.
B. Limit the text to phrases, not complete sentences.
C. Create one heading for each slide or image.
D. Keep the backgrounds simple so they do not conflict with text.

_____ 13. Which of the following items is a peripheral computer device that allows photographs, slides, transparencies, diagrams, or illustrations to be digitized for integration into a presentation, into a document, or for storage on a CD-ROM or DVD?

A. Camera
B. Scanner
C. Video capture device
D. CD/DVD burner

_____ 14. Which of the following equipment captures images on film or digitally in electronic memory?

A. Camera
B. Scanner
C. Video capture device
D. CD/DVD burner

_____ 15. Which of the following equipment can take an individual frame or short segment of a video and digitize it for use in a document or presentation?

A. Camera
B. Scanner
C. Video capture device
D. CD/DVD burner

True / False

B. Write the correct letters on the blanks.

_____ 1. Learning objectives can be very broad and general in nature.

A. True
B. False

_____ 2. One guideline for developing effective visual aids is to use a maximum of 10 lines down and 10 words across the viewing area.

A. True
B. False

_____ 3. A lesson plan provides uniformity by standardizing the instruction and enabling instructors to provide the same information in a similar format each time the lesson is taught.

A. True
B. False

_____ 4. A lesson plan states clearly what an instructor will accomplish with students during a particular lesson.

A. True
B. False

_____ 5. The learning objective model developed by Mager consists of two components: the audience and the performance statement.

A. True
B. False

_____ 6. Learning objectives provide a foundation for instructional design and aid in overall course development.

A. True
B. False

_____ 7. Learning objectives need to be specific and measurable.

A. True
B. False

_____ 8. Instructors must use the detailed outline format when writing lesson plans.

A. True
B. False

_____ 9. In the Mager model for writing learning objectives, the performance statement identifies what the student is expected to do.

A. True
B. False

_____ 10. The first step in creating a lesson plan is to determine the order for instructing the material.

A. True
B. False

_____ 11. Even when a majority of students fail to perform satisfactorily, the lesson plan should NOT be revised.

A. True
B. False

_____ 12. When it is determined that there is a problem with a lesson plan, that lesson plan should be totally discontinued from further use.

A. True
B. False

_____ 13. When creating a lesson plan, the instructor should develop lesson activities and identify and develop instructional aids to support instruction.

A. True
B. False

_____ 14. The lesson plan format in which only major statements or words are included is best used for experienced instructors with a wealth of knowledge regarding the topic.

A. True
B. False

_____ 15. Components in a lesson plan may need to be modified to include current information or changes in operating policies and procedures.

A. True
B. False

Describe

C. Write the components of a learning objective and their descriptions according to the Mager model on the lines provided.

1. ______________________________

____

____

2. ______________________________

____

____

3. ______________________________

____

____

Chapter 15 Instructor and Course Evaluations

Multiple Choice

A. Write the correct letters on the blanks.

_____ 1. Which of the following is the ongoing, repeated checking during course development and during instruction to determine the most effective instructional content, presentation methods, training aids, and testing techniques?

A. Formative evaluation
B. Summative evaluation
C. Process evaluation
D. Product evaluationn

_____ 2. Which of the following is NOT a guideline for instructor evaluations?

A. Maintain consistency.
B. State criteria clearly.
C. Maintain subjectivity.
D. Train supervisors properly.

_____ 3. Which of the following refers to the process of teaching the course on a trial basis?

A. Field test
B. Observation
C. Initial run
D. Course rehearsal

_____ 4. Which of the following is an end-of-the-course appraisal that commonly measures learning by some form of objective or subjective evaluation instrument?

A. Formative evaluation
B. Summative evaluation
C. Process evaluation
D. Product evaluation

_____ 5. Which of the following is NOT a characteristic of a formal evaluation?

A. Occurs daily
B. Always held in private
C. May occur annually
D. Uses results of student instructor evaluation survey forms

_____ 6. Which of the following statements regarding guidelines for instructor evaluations after the evaluation has been performed is TRUE?

A. Discuss the evaluation in front of all staff.
B. Do NOT ask for the instructor's opinion on individual performance.
C. When the evaluation is discouraging, suggest the instructor look for other work.
D. Explain what the instructor did well and what areas may need improvement.

_____ 7. Which of the following statements regarding guidelines for instructor evaluations during the evaluation is TRUE?

A. Do NOT take notes during the evaluation.
B. Evaluate instructors based partially upon rumors.
C. Use the evaluation form provided by the organization.
D. Do NOT provide feedback during the evaluation process.

_____ 8. Which of the following statements about survey forms is TRUE?

A. Students should be required to sign the survey form.
B. Surveys should NOT include open-ended questions.
C. Surveys should NOT ask questions requiring a personal opinion.
D. Surveys often are answered on a continuum from very satisfied to very dissatisfied.

_____ 9. Which of the following is the first step to take after reviewing evaluation results?

A. Identify actions to correct deficiencies.
B. Determine causes for student failure.
C. Document and report results to superiors.
D. Determine whether reviewing the results is warranted.

_____ 10. Which of the following is NOT a point that would be included on instructor evaluation forms?

A. Appropriate cost for certification
B. Frequent and appropriate use of questions
C. Use of appropriate instructional methods
D. Interaction between the instructor and students

True / False

B. Write the correct letters on the blanks.

_____ 1. The strengths and weaknesses of an employee that are discussed during a personnel evaluation do NOT become part of a permanent record.
A. True
B. False

_____ 2. One benefit of a personnel evaluation program is that the instructor or supervisor who is performing the evaluation becomes more familiar with the instructors being evaluated.
A. True
B. False

_____ 3. When critiquing instructor performance either informally or formally, the instructor-supervisor should first provide any negative comments.
A. True
B. False

_____ 4. When conducting formal instructor evaluations, the instructor-supervisor should NOT discuss the evaluation process with the instructor being evaluated.
A. True
B. False

_____ 5. If an informal evaluation indicates the instructor is performing below expectations, comments and suggestions for improvement should be done in private.
A. True
B. False

_____ 6. The subordinate being evaluated should NOT be included in the decision-making process used to apply the strengths constructively or correct the weaknesses identified in an evaluation.
A. True
B. False

_____ 7. Individual employee strengths identified in an evaluation should be used to the benefit of the individual instructor, division, organization, and community.
A. True
B. False

_____ 8. The subordinate should be permitted to determine the steps necessary to overcome a weakness identified in an evaluation.
A. True
B. False

_____ 9. Students can be sent home with surveys and will generally return them promptly.
A. True
B. False

_____ 10. Students should be required to sign their names on evaluations.
A. True
B. False

_____ 11. Student surveys of instructor performance should include questions on preparation, presentation skills, and interaction with students.
A. True
B. False

_____ 12. A summative evaluation looks at the product and evaluates reactions to the course and instructional methods.
A. True
B. False

_____ 13. A formative evaluation looks at the process of course development and instruction.
A. True
B. False

_____ 14. The training course/lesson evaluation process should also include surveys of supervisory personnel such as incident commanders (ICs), supervisors, and division managers within the organization.
A. True
B. False

_____ 15. When using observation as a method for conducting formative evaluations, the evaluation is a static, one-time event that occurs at the end of an instructional period.
A. True
B. False

Describe

C. Write the questions that a thorough course/lesson evaluation should answer for the following areas on the lines provided.

1. Reaction ______________________________

2. Knowledge ______________________________

3. Skills ______________________________

4. Attitudes ______________________________

5. Transfer of learning ______________________________

6. Results ______________________________

15

Terms

D. Write the meaning of the following terms on the lines provided.

1. Formative evaluation ______________________________

____

____

2. Summative evaluation ______________________________

____

____

3. Field test ______________________________

____

____

Chapter 16 Student Testing Instruments

Multiple Choice

A. Write the correct letters on the blanks.

_____ 1. Which of the following tests rate student performance compared to other students based on broad sampling?

A. All test types
B. Formative tests
C. Norm-referenced tests
D. Criterion-referenced tests

_____ 2. Which of the following tests compare performance against appropriate minimum standards?

A. All test types
B. Formative tests
C. Norm-referenced tests
D. Criterion-referenced tests

_____ 3. Which of the following must tests be based on?

A. Class quizzes
B. Test banks
C. Program goals
D. Learning objectives

_____ 4. Which of the following tests would be used if the test needs to measure manipulative skills in the psychomotor domain?

A. Oral test
B. Written test
C. Formative test
D. Performance or skill test

_____ 5. Which of the following tests is designed to determine readiness for instruction or placement in the appropriate instructional level?

A. Diagnostic test
B. Formative or progress test
C. Prescriptive or placement test
D. Summative or comprehensive test

_____ 6. Which of the following tests is designed to rate terminal performance?

A. Diagnostic test
B. Formative or progress test
C. Prescriptive or placement test
D. Summative or comprehensive test

_____ 7. Which of the following tests is designed to measure improved progress or identify learning problems that are hampering progress?

A. Diagnostic test
B. Formative or progress test
C. Prescriptive or placement test
D. Summative or comprehensive test

_____ 8. Which of the following tests is designed to determine student-learning difficulties?

A. Diagnostic test
B. Formative or progress test
C. Prescriptive or placement test
D. Summative or comprehensive test

_____ 9. Which of the following is a purpose of test specifications?

A. Provide documentation for legal purposes
B. Ensure that all students can easily pass the test
C. Eliminate the time and effort instructors put into creating tests
D. Ensure that a representative sampling of questions are created to determine student understanding of the learning objectives and course outcomes

_____ 10. Which of the following is a function of a test planning sheet?

A. To eliminate the time instructors spend creating a test
B. To determine whether the instructor is successful in teaching
C. To specify which students will succeed and which students will fail
D. To specify the number of test items for each topic and learning objective

_____ 11. Which of the following would NOT be a language and comprehension barrier to test taking?

A. Vague directions
B. Directions under each subsection
C. Lengthy, complex, or unclear sentences
D. Higher reading level than the student audience possesses

_____ 12. Which of the following would be a factor that may give clues to test answers?

A. Answers placed in different locations
B. Word associations that give away the answer
C. Answers that are used more than once
D. Answers that are scientific or technical in nature

_____ 13. Which of the following is NOT a factor in determining the number of test items?

A. Purpose of the test
B. Types of test items or performance items
C. Cost of printing or copying the written test
D. Reliability level of the test to accurately determine student abilities

_____ 14. Which of the following is NOT a characteristic that ensures test usability?

A. Cost-effective
B. Difficult to take
C. Easy to administer
D. Sufficient time for administration

_____ 15. Which of the following is the most significant consideration for test design?

A. The majority of students must be able to pass.
B. All test items must be capable of being electronically scored.
C. Tests must include a variety of testing formats.
D. All test items must be referenced to a learning objective.

_____ 16. Which of the following refers to tests that assess the student's ability to perform a test within a specific time?

A. Speeded test
B. Powered test
C. Monitored test
D. Criterion-timed test

_____ 17. Which of the following is the recommended arrangement for test items?

A. In a random order
B. In a sequence that fits the page layout needed
C. In a sequence of decreasing difficulty from complex to simple
D. In a sequence of increasing difficulty from simple to complex

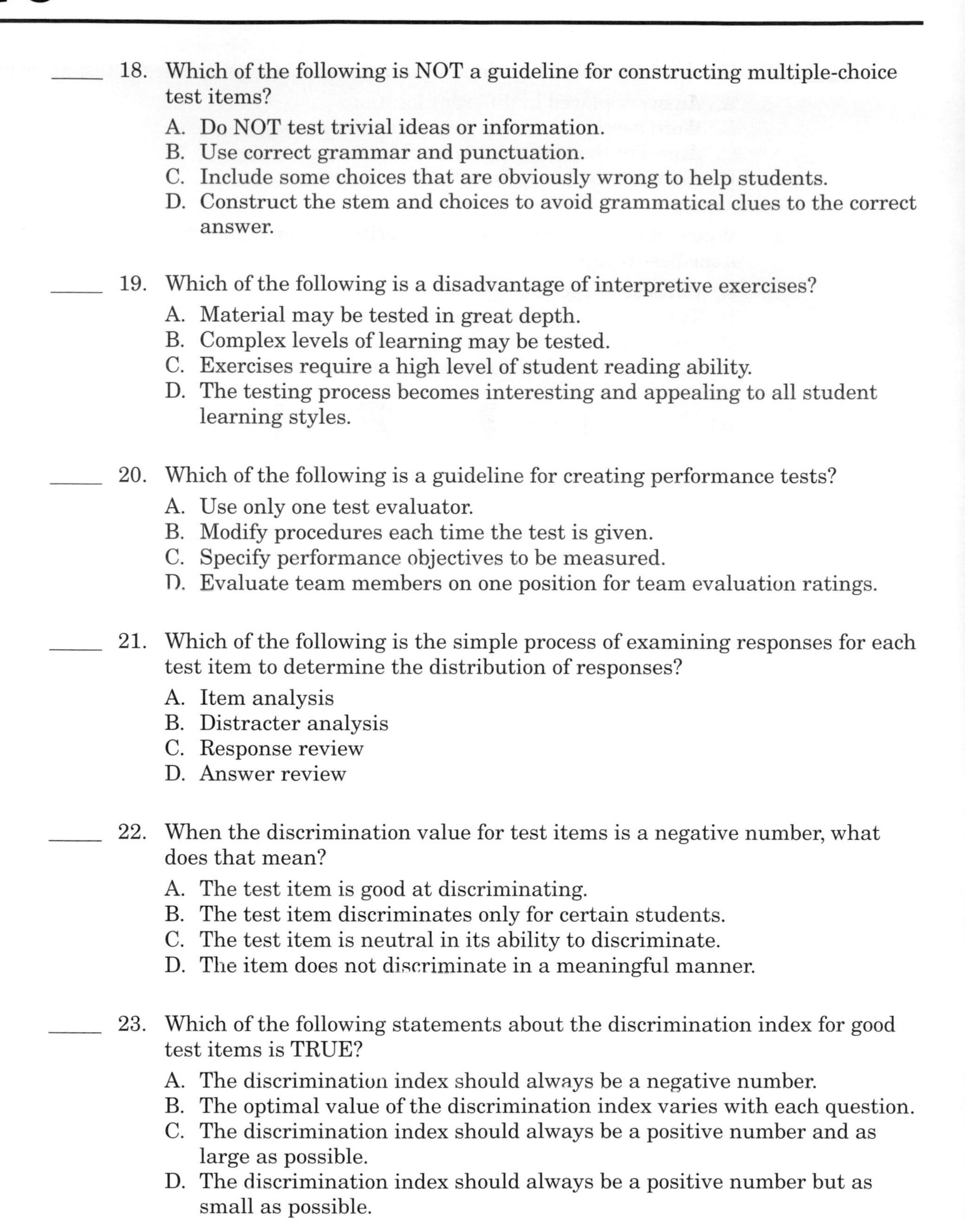

_____ 18. Which of the following is NOT a guideline for constructing multiple-choice test items?

A. Do NOT test trivial ideas or information.
B. Use correct grammar and punctuation.
C. Include some choices that are obviously wrong to help students.
D. Construct the stem and choices to avoid grammatical clues to the correct answer.

_____ 19. Which of the following is a disadvantage of interpretive exercises?

A. Material may be tested in great depth.
B. Complex levels of learning may be tested.
C. Exercises require a high level of student reading ability.
D. The testing process becomes interesting and appealing to all student learning styles.

_____ 20. Which of the following is a guideline for creating performance tests?

A. Use only one test evaluator.
B. Modify procedures each time the test is given.
C. Specify performance objectives to be measured.
D. Evaluate team members on one position for team evaluation ratings.

_____ 21. Which of the following is the simple process of examining responses for each test item to determine the distribution of responses?

A. Item analysis
B. Distracter analysis
C. Response review
D. Answer review

_____ 22. When the discrimination value for test items is a negative number, what does that mean?

A. The test item is good at discriminating.
B. The test item discriminates only for certain students.
C. The test item is neutral in its ability to discriminate.
D. The item does not discriminate in a meaningful manner.

_____ 23. Which of the following statements about the discrimination index for good test items is TRUE?

A. The discrimination index should always be a negative number.
B. The optimal value of the discrimination index varies with each question.
C. The discrimination index should always be a positive number and as large as possible.
D. The discrimination index should always be a positive number but as small as possible.

_____ 24. Which of the following refers to the average score?
A. Mean
B. Median
C. Mode
D. Middle

_____ 25. When using a difficulty index with criterion-referenced tests, what number is the best result?
A. 0.5
B. 0.7
C. 0.9
D. 1.0

True / False

B. Write the correct letters on the blanks.

_____ 1. The majority of tests written for fire and emergency services personnel are criterion-based.
A. True
B. False

_____ 2. The purpose of a test depends partly on the learning domain that is being tested.
A. True
B. False

_____ 3. The first step in test planning is to construct appropriate test items.
A. True
B. False

_____ 4. Test questions are specifically written to address lesson objectives.
A. True
B. False

_____ 5. Learning objectives are generally created in a hierarchal format with the course or learning outcomes at the bottom stated in specific terms.
A. True
B. False

_____ 6. Criterion-referenced tests determine how well the individual student has achieved the learning objectives established for the lesson or course.
A. True
B. False

_____ 7. Norm-referenced tests rank the members of the class.
A. True
B. False

_____ 8. Reliability is the extent to which a test measures what it is supposed to measure.
A. True
B. False

_____ 9. The three basic types of test instruments are written tests, oral tests, and performance tests.
A. True
B. False

_____ 10. Two approaches to determining the number of test questions are the guideline approach and the mathematical approach.
A. True
B. False

_____ 11. Validity is the consistency and accuracy of test measurement.
A. True
B. False

_____ 12. When creating multiple-choice test items write the stem in the form of a direct question or an incomplete sentence that asks and measures only one learning objective.
A. True
B. False

_____ 13. A subjective test item has only one correct answer.
A. True
B. False

_____ 14. When creating multiple-choice test items, place the correct answers in consistent positions among the A, B, C, and D choices.
A. True
B. False

_____ 15. When creating true-false test items include a greater number of false items.
A. True
B. False

_____ 16. A major advantage of multiple-choice test items is the ability to measure achievement, complex learning objectives, and various types of knowledge.

A. True
B. False

_____ 17. When creating true-false test items avoid creating test items that could trick or mislead students into making a mistake.

A. True
B. False

_____ 18. A disadvantage of true-false test items is that students can guess the item correctly without knowing anything about the subject matter.

A. True
B. False

_____ 19. When creating matching test items arrange responses randomly.

A. True
B. False

_____ 20. Matching tests are considered inferior to multiple-choice tests in measuring high levels of instruction.

A. True
B. False

_____ 21. An advantage of matching test items is that they can cover a large amount of factual material in a compact space.

A. True
B. False

_____ 22. A guideline for creating short-answer/completion test items is to arrange the statement in order to place the blanks at the beginning of the sentence.

A. True
B. False

_____ 23. One of the most important advantages of short-answer/completion test items is that students must supply the answer, which minimizes the possibility that they can guess the correct answer.

A. True
B. False

_____ 24. An advantage of short-answer/completion test items is that they are easy to score.
A. True
B. False

_____ 25. When creating an essay test item give thorough and specific directions that designate the time to be spent on each question or the length of each response.
A. True
B. False

_____ 26. A disadvantage of essay test items is that instructors may find it difficult to read the students' response, which affects the ability to score essays accurately.
A. True
B. False

_____ 27. A disadvantage of essay test items is that they are difficult to create and cannot test higher learning levels.
A. True
B. False

_____ 28. A guideline for creating an interpretive exercise is to use a variety of test item types in sufficient quantity to provide a good sample of a student's ability to interpret the introductory material.
A. True
B. False

_____ 29. Oral tests are commonly used in the fire and emergency services.
A. True
B. False

_____ 30. When the purpose of an oral test is to determine knowledge, then the questions should be open.
A. True
B. False

_____ 31. An oral test is the only valid method of measuring students' ability to verbally communicate ideas, concepts, or processes.
A. True
B. False

_____ 32. A performance test is the only valid method of measuring a student's achievement and ability to perform manipulative skills.

A. True
B. False

_____ 33. A disadvantage of performance tests is that they are both time- and resource-consuming.

A. True
B. False

_____ 34. Grading is the act of identifying which answers are right and which are wrong.

A. True
B. False

_____ 35. Scoring is the act of assigning a value to the score.

A. True
B. False

_____ 36. Using the criterion-referenced grading system always results in the majority of students passing the lesson or course.

A. True
B. False

_____ 37. The basis of scoring and assigning grades should be a composite (mixture) of various lesson/course activities and other factors.

A. True
B. False

_____ 38. Instructors are NOT obligated to defend any score or grade given.

A. True
B. False

_____ 39. When an entire class scores poorly or the scores are skewed toward the lower end of the scale, the instructor should perform a careful analysis to determine the reasons for poor testing performance.

A. True
B. False

_____ 40. One corrective technique for skewed test results is to throw out the poor items and recalculate the score.

A. True
B. False

List

C. Write the steps in test planning on the lines provided.

1. Step 1:____
2. Step 2:____
3. Step 3:____
4. Step 4:____

D. Write the three basic types of test instruments on the lines provided.

1. ____
2. ____
3. ____

Describe

E. Describe an interpretive exercise on the lines provided.

Chapter 17 Course and Evolution Management

Multiple Choice

A. Write the correct letters on the blanks.

_____ 1. Which of the following is an advantage of discussions?

A. Works with certain topics only
B. Encourages critical thinking among students
C. Depends on student abilities and knowledge to be successful
D. Discourages shy or withdrawn students to participate

_____ 2. Which of the following is a disadvantage of discussions?

A. Requires less instructor preparation
B. Promotes anxiety and nervousness in students
C. Provides an informal assessment of student progress
D. Becomes student centered by placing focus and responsibility on students

_____ 3. Which of the following techniques allows students to experience the process used in organizational decision-making and requires that ideas be realistic?

A. Debate
B. Brainstorm
C. Nominal group process
D. Agenda-based process

_____ 4. Which of the following techniques involves selecting a controversial topic and asking students to discuss the pros and cons of the topic?

A. Debate
B. Brainstorm
C. Nominal group process
D. Agenda-based process

_____ 5. Which of the following techniques involves accepting all ideas and then discussing the relative merits of each?

A. Debate
B. Brainstorm
C. Nominal group process
D. Agenda-based process

_____ 6. Which of the following techniques involves creating an agenda of topics or key points and providing it to students?

A. Debate
B. Brainstorm
C. Nominal group process
D. Agenda-based process

_____ 7. Which of the following ICS functions ensures the safety of all students, advises the instructor on safety-related issues, and has the authority to immediately halt any unsafe act or practice?

A. Staging officer
B. Communications officer
C. Incident safety officer
D. Division/group/branch officer

_____ 8. Which of the following ICS functions manages the staging area for multiple company/agency evolutions?

A. Staging officer
B. Communications officer
C. Incident safety officer
D. Division/group/branch officer

_____ 9. Which of the following ICS functions provides the materials and supplies required to perform the evolutions?

A. Staging officer
B. Communications officer
C. Logistics officer
D. Division/group/branch officer

_____ 10. Which of the following ICS functions manages all incident operations and is primarily responsible for formulating the incident action plan?

A. Staging officer
B. Incident commander
C. Incident safety officer
D. Division/group/branch officer

_____ 11. Which of the following ICS functions manages or supervises students in the various locations around the site, including classroom or assembly area where students receive final instructions, briefings, or postincident critiques?

A. Staging officer
B. Incident commander
C. Incident safety officer
D. Division/group/branch officer

_____ 12. Where is the incident action plan maintained during a practical training evolution?

A. At the fire station
B. At the incident command post
C. With the battalion chief
D. In the most convenient location

_____ 13. Which of the following contains all tactical and support activities required for the control of the training?

A. Incident action plan
B. Standard operating guidelines
C. Incident objectives
D. Post-incident critique

_____ 14. For training purposes, when should the ICS model be used?

A. At training evolutions where the media may be present
B. ICS does NOT need to be used for training evolutions
C. Only at training evolutions involving live-fire or high-risk
D. At all types of training evolutions whether they involve live fire or not

_____ 15. Which of the following is NOT a purpose of a postincident critique of a practical training evolution?

A. Evaluate student skills and learning
B. Determine whether the training went over budget
C. Determine safety problems that need to be corrected
D. Determine the effectiveness of the organization's ICS model

True / False

B. Write the correct letters on the blanks.

_____ 1. Discussions should NOT be announced in advance to students.

A. True
B. False

_____ 2. When an instructor decides that a discussion is appropriate for a class, it is important to start at a fast pace to keep up interest.

A. True
B. False

_____ 3. The result of a discussion may be a solution to a problem, a decision, or the increase in personal knowledge.

A. True
B. False

_____ 4. A disadvantage of discussions is that they build interdependence between students and encourage teamwork.
A. True
B. False

_____ 5. When preparing for a classroom discussion, the instructor should explain rules regarding interpersonal relations such as how students who wish to speak are recognized.
A. True
B. False

_____ 6. In small group discussions, the instructor is part of the group.
A. True
B. False

_____ 7. In large group discussions, the group votes on a topic that will be discussed.
A. True
B. False

_____ 8. Small group discussions work best when students are experienced in working with others.
A. True
B. False

_____ 9. Both large and small group discussions require leadership on the part of the instructor or group facilitator.
A. True
B. False

_____ 10. When leading a discussion, the role of the director is to ensure that all students have an opportunity to speak and no one dominates the discussion.
A. True
B. False

_____ 11. The responsibility of being familiar with and implementing the procedures of the ICS adopted by the local jurisdiction is beyond that of the Level II instructor.
A. True
B. False

_____ 12. ICS is NOT officially a part of the National Incident Management System (NIMS) and its use is optional by federal agencies or agencies that receive federal funding.
A. True
B. False

_____ 13. One benefit of the use of ICS during practical training evolutions is that it ensures the safety and accountability of students.
A. True
B. False

_____ 14. Components of an incident command system include common terminology, unified command structure, and consolidated action plans.
A. True
B. False

_____ 15. When planning a multiagency training evolution, training should include the use of similar equipment, terminology, procedures, and processes.
A. True
B. False

List

C. Write incident command system (ICS) components on the lines provided.

1. ____________________
2. ____________________
3. ____________________
4. ____________________
5. ____________________
6. ____________________
7. ____________________
8. ____________________

Terms

D. Write the meaning of the following terms on the lines provided.

1. IAP ______________________________
2. ICS ______________________________
3. NIMS ______________________________
4. ISO ______________________________

Chapter 18 Administrative Duties

Multiple Choice

A. Write the correct letters on the blanks.

_____ 1. Which of the following is the first step of the data-gathering process?
   A. List sources for each information type.
   B. Eliminate topics difficult to research.
   C. Locate funds that can be used in the research.
   D. Identify the topic that is to be researched.

_____ 2. Which of the following is NOT a factor in determining who will perform the research?
   A. Time available
   B. Importance of the project
   C. Promotional opportunity
   D. Funds available to hire a specialist

_____ 3. Which of the following is literature that has been twice removed from the original source and may contain errors of translation, interpretation, or context?
   A. Primary literature
   B. Secondary literature
   C. Tertiary literature
   D. Incidental literature

_____ 4. Which of the following statements about using the Internet as a research tool is TRUE?
   A. Download time is the same for all web sites.
   B. Information on the Internet is always free of charge.
   C. All information found on the Internet is reliable.
   D. Information found on the Internet must be verified for accuracy.

_____ 5. Which of the following testing and standards organizations maintains data on fire-related loss, injuries, deaths, fire causes, and other fire- and safety-related topics and also creates consensus standards?
   A. FM Global
   B. Congressional Fire Services Institute
   C. Underwriters Laboratories, Inc
   D. National Fire Protection Association

_____ 6. Which of the following is NOT a characteristic of valid data or sources?
A. Subjectivity
B. Credibility
C. Accuracy
D. Support

_____ 7. Which of the following types of analyses is based on the relationship between the effort and the result?
A. Process analysis
B. Policy analysis
C. Program analysis
D. Cost/benefit analysis

_____ 8. Which of the following types of analyses determines the physical activities that members of the organization must perform as part of their assigned duties?
A. Process analysis
B. Task analysis
C. Program analysis
D. Cost/benefit analysis

_____ 9. Which of the following types of analyses determines the likelihood of an event occurring such as a major fire or natural disaster?
A. Needs analysis
B. Risk analysis
C. Process analysis
D. Program analysis

_____ 10. Which of the following types of analyses determines the types of services that an organization is currently delivering and comparing them to the services that will be needed in the future?
A. Needs analysis
B. Risk analysis
C. Process analysis
D. Program analysis

_____ 11. Which of the following types of analyses determines the most efficient ways to provide a program or service by looking at each program component?
A. Needs analysis
B. Risk analysis
C. Program analysis
D. Cost/benefit analysis

_____ 12. Which of the following is a characteristic of a capital budget?

A. Funded from general revenue sources
B. Funded by one-time, earmarked funds
C. Used to pay for the recurring expenses of day-to-day operations
D. Generally developed over a period of 6 months before adoption of a new budget

_____ 13. Which of the following is a characteristic of an operating budget?

A. Includes projected major purchases
B. Funded by one-time, earmarked funds
C. Used to pay for the recurring expenses of day-to-day operations
D. Involve a longer development time and may require many years for completion

_____ 14. Which of the following steps in the budget development process is part of the strategic planning process?

A. Step 1: Plan
B. Step 2: Prepare
C. Step 3: Implement
D. Step 4: Monitor

_____ 15. In which of the following steps in the budget development process are estimated revenues from all sources translated into preliminary budget priorities by the finance and revenue department of the jurisdiction?

A. Step 1: Plan
B. Step 2: Prepare
C. Step 3: Implement
D. Step 4: Monitor

_____ 16. In which of the following steps in the budget development process can the results be used to prevent a budgetary crisis in the event of a change in the economic environment?

A. Step 3: Implement
B. Step 4: Monitor
C. Step 5: Evaluate
D. Step 6: Revise

_____ 17. In which of the following steps in the budget development process are the budgeted funds used for programs and activities that provide the services approved by the jurisdiction?

A. Step 1: Plan
B. Step 2: Prepare
C. Step 3: Implement
D. Step 4: Monitor

_____ 18. In which step in the purchasing process should criteria be established and points assigned based on the equipment's ability to meet the standard or regulation?
A. Step 1: Determine the needs of the department/organization.
B. Step 2: Conduct research on the equipment, manufacturers, and any applicable standards/regulations.
C. Step 3: Evaluate and field-test proposed equipment.
D. Step 4: Review product data.

_____ 19. Which of the following steps in the purchasing process may require a formal bid process?
A. Step 3: Evaluate and field-test proposed equipment.
B. Step 4: Review product data.
C. Step 5: Conduct the purchasing process.
D. Step 6: Evaluate purchasing process and revise when necessary.

_____ 20. In which of the following steps in the purchasing process would units not meeting the criteria be eliminated from consideration?
A. Step 3: Evaluate and field-test proposed equipment.
B. Step 4: Review product data.
C. Step 5: Conduct the purchasing process.
D. Step 6: Evaluate purchasing process and revise when necessary.

True / False

B. Write the correct letters on the blanks.

_____ 1. Information obtained from vendors/manufacturers of products does NOT need to be compared to any other information.
A. True
B. False

_____ 2. The U.S. government maintains data of fire and emergency responses through the National Fire Incident Reporting System (NFIRS) and on fire equipment failures through the National Institute for Occupational Safety and Health (NIOSH).
A. True
B. False

_____ 3. Sources of information for the fire and emergency services instructor include the Internet, government agencies, libraries, educational institutions, and professional organizations.
A. True
B. False

_____ 4. Instructors do NOT need to worry about citations because the material is for educational purposes.
A. True
B. False

_____ 5. In the analysis process the best approach is to develop only one very sound approach to a problem.
A. True
B. False

_____ 6. Canadian and U.S. national government agencies at all levels are the best sources of raw data on fire and emergency services topics.
A. True
B. False

_____ 7. Developing and managing budgets are routine activities that can be completed by any member of the organization.
A. True
B. False

_____ 8. One function of a budget for a governmental jurisdiction is to provide assistance in the decision-making process.
A. True
B. False

_____ 9. Funds donated through grants/gifts for capital purchases can be used for operating expenses, if necessary.
A. True
B. False

_____ 10. The majority of governmental jurisdictions depend on revenue from property, sales, or income taxes or a combination of these taxes as the primary source of revenue.
A. True
B. False

_____ 11. Fundraising is most often an activity of volunteer or combination emergency services organizations that must supplement or provide their own operating revenue.
A. True
B. False

_____ 12. An effective purchasing process must be a subjective process and take into account emotions.
A. True
B. False

_____ 13. When conducting a review of product data for the purchasing process, units that do NOT meet criteria should be placed at the bottom of the list for consideration.
A. True
B. False

_____ 14. An effective purchasing process must be repeatable by future personnel who are given the task of providing logistical support through the process.
A. True
B. False

_____ 15. The purchasing procedure for fire protection equipment depends on the process adopted and regulated by the authority having jurisdiction (AHJ).
A. True
B. False

List

C. Write sources of information for the fire and emergency services instructor on the lines provided.

1. ____________________
2. ____________________
3. ____________________
4. ____________________
5. ____________________
6. ____________________
7. ____________________
8. ____________________

Terms

D. Write the meaning of the following terms on the lines provided.

1. Search engines ______________________________

____

____

2. Capital budget ______________________________

____

____

3. Operating budget______________________________

____

____

Chapter 19 Supervision and Management

Multiple Choice

A. Write the correct letters on the blanks.

_____ 1. Which of the following functions of the management process involves setting goals and objectives and determining the direction the organization or unit will take to achieve those results?

A. Planning
B. Organizing
C. Leading
D. Controlling

_____ 2. Which of the following is NOT a characteristic of objectives for a workgroup?

A. Attainable
B. Measurable
C. Clearly stated
D. Outside of the capability of the workgroup

_____ 3. Which of the following functions of the management process involves establishing and implementing the mechanisms to ensure that objectives are attained?

A. Planning
B. Organizing
C. Leading
D. Controlling

_____ 4. Which of the following is NOT a major responsibility of a supervisor?

A. Maintain discipline.
B. Maintain files and records and prepare reports.
C. Set a clear and positive example for subordinates.
D. Determine agency-wide policies and procedures.

_____ 5. Which of the following functions of the management process involves influencing, inspiring, and motivating employees to achieve goals and objectives?

A. Planning
B. Organizing
C. Leading
D. Controlling

_____ 6. Which of the following is NOT a key element that must be part of the instructor's supervisory style?

A. Keep accurate records.
B. Delegate or involve team members in planning.
C. Show favoritism to team members who accomplish more.
D. Show consideration for diversity within the workgroup.

_____ 7. Which of the following functions of the management process involves coordinating tasks and resources to accomplish the goals and objectives?

A. Planning
B. Organizing
C. Leading
D. Controlling

_____ 8. Which of the following is the first step taken when creating a training schedule?

A. Determine needs
B. Coordinate training
C. Publish the schedule
D. Revise the schedule

_____ 9. When should availability of instructors and facilities be determined?

A. Whenever convenient
B. After creating the training schedule
C. Before creating the training schedule
D. Either before or after creating the training schedule

_____ 10. Which of the following is the final step in establishing a training schedule?

A. Revise the schedule.
B. Publish and make available the schedule.
C. Coordinate training with other jurisdictions.
D. Determine the availability of instructors.

True / False

B. Write the correct letters on the blanks.

_____ 1. When the first sign of a problem arises, such as excessive griping, the supervisor should wait and see if the problem goes away by itself.

A. True
B. False

_____ 2. When a problem is very complex or outside the skills of the supervisor, the instructor needs to refer the employee to the organization's employee assistance program or human resources department.

A. True
B. False

_____ 3. When a personnel incident does occur, a complete record of the incident and the counseling session should be kept.

A. True
B. False

_____ 4. In order to create an effective team, ensure that teams are composed of people from similar backgrounds.

A. True
B. False

_____ 5. One method of involving employees in the process of establishing goals and objectives is to delegate tasks.

A. True
B. False

_____ 6. In order to create an effective team, the supervisor should establish measurements for success and determine whether the team and team members are successful.

A. True
B. False

_____ 7. Two ways to create job interest within a team are to reward employees and celebrate accomplishments.

A. True
B. False

_____ 8. When completing tasks, final responsibility and authority must always remain with the employee.

A. True
B. False

_____ 9. When completing tasks, employees should be given minimal amounts of responsibility and authority on a project.

A. True
B. False

_____ 10. Ensuring teamwork and cooperation within the organization are responsibilities of the supervisor.

A. True
B. False

_____ 11. An employee in the fire and emergency services generally performs the same task all the time.

A. True
B. False

_____ 12. All supervisors are responsible for the maintenance of records and reports such as daily timesheets and performance evaluations.

A. True
B. False

_____ 13. Resources that the instructor has to achieve goals and objectives include the following: human, financial, physical, information, and time.

A. True
B. False

_____ 14. Management skills required to be an effective manager include technical skills, interpersonal communication skills, and conceptual and decision-making skills.

A. True
B. False

_____ 15. The Level II instructor does NOT have the responsibility of scheduling resources and instructional delivery.

A. True
B. False

_____ 16. The final step taken when creating a training schedule is to revise the schedule.

A. True
B. False

_____ 17. Student availability is NOT a factor when scheduling training sessions.
A. True
B. False

_____ 18. Factors that influence scheduling include governmental mandates, instructor availability, and funds.
A. True
B. False

_____ 19. Governmental mandates rarely change over time.
A. True
B. False

_____ 20. Factors in determining projected training needs include increases in service levels and expansion of coverage areas.
A. True
B. False

List

C. Write the three ways that an instructor can create job interest within a team on the lines provided.

1. ____________________

2. ____________________

3. ____________________

D. **Write the nine major responsibilities of a supervisor on the lines provided.**

1. ______________________________

____

2. ______________________________

____

3. ______________________________

____

4. ______________________________

____

5. ______________________________

____

6. ______________________________

____

7. ______________________________

____

8. ______________________________

____

9. ______________________________

____

Chapter 20 Administration: Records, Policies, and Personnel

Multiple Choice

A. Write the correct letters on the blanks.

_____ 1. Which of the following components of a records-management system consists of system support that ensures the system continues to operate correctly and efficiently?

A. Policies
B. Procedures
C. Maintenance
D. Technology

_____ 2. Which of the following components of a records-management system consists of codified statements that define the system, the data to be gathered, and how data are stored, accessed, analyzed, and disposed of?

A. Policies
B. Procedures
C. Tools
D. Ongoing supporting education

_____ 3. Which of the following is NOT information that would be gathered for a training records system?

A. Student attendance rosters
B. Topics taught at each session
C. Income and work history of students
D. Dates of each training session

_____ 4. Which of the following prohibits the release of an individual's evaluation/testing scores in the U.S.?

A. Education Reform Bill
B. Right-to-Know Act
C. Family Education and Privacy Act
D. Education and Testing Regulation Act

_____ 5. Which of the following is a guiding principle or rule that organizations develop, adopt, and use as a basis or foundation for decision-making?

A. Policy
B. Procedure
C. Guideline
D. Recommendation

_____ 6. Which of the following identifies a general philosophy and provides direction with latitude for achieving the overall goal?
A. Policy
B. Procedure
C. Guideline
D. Recommendation

_____ 7. Which of the following identifies the steps that must be taken to fulfill the intent of a policy and is written to support a policy?
A. Outline
B. Procedure
C. Guideline
D. Recommendation

_____ 8. Which of the following is the first step taken when determining the need for a new policy, procedure, or guideline?
A. Collect the data to evaluate the need.
B. Select the evaluation model.
C. Establish a revision process or schedule.
D. Identify the problem.

_____ 9. Which of the following is the first step in the adoption process for a policy, procedure, or guideline?
A. Identify a need.
B. Develop a draft document.
C. Adopt the document.
D. Implement the document's contents.

_____ 10. In which step of the adoption process for a policy, procedure, or guideline should personnel be provided an opportunity to respond with feedback and input on the document?
A. Adopt the document.
B. Develop a draft document.
C. Submit the draft for organizational review.
D. Implement the document's contents.

_____ 11. Which of the following is the final step in the adoption process for a policy, procedure, or guideline?
A. Adopt the document.
B. Develop a draft document.
C. Evaluate its effectiveness.
D. Implement the document's contents.

_____ 12. Which of the following steps, if not done properly, is the primary cause for the failure of personnel to accept and adhere to policies, procedures, and guidelines?

A. Adopt the document.
B. Develop a draft document.
C. Evaluate its effectiveness.
D. Implement the document's contents.

_____ 13. Which of the following is a guideline for personnel evaluations?

A. State goals and objectives clearly and concisely.
B. Have higher standards for employees who are performing well.
C. Surprise employees with evaluation times.
D. Conduct evaluations only if employees are performing below expectations.

_____ 14. Which of the following statements regarding responses for a 360-degree feedback evaluation is TRUE?

A. Employees should be mandated to provide responses.
B. Responses should become available to all staff members.
C. Responses that are neutral should not be considered.
D. Responses must remain confidential to protect those providing information.

_____ 15. Which of the following statements regarding the evaluation process is TRUE?

A. Supervisors can create their own guidelines for evaluations.
B. Evaluations may or may not be supported by documentation.
C. Evaluations must be supported by definite identifiable criteria.
D. Evaluations can be subjective and based on subjective observations.

True / False

B. Write the correct letters on the blanks.

_____ 1. The raw data stored in records can be used to support recommendations.

A. True
B. False

_____ 2. Records management includes the planning, controlling, directing, organizing, training, and conducting of other managerial activities required for maintaining an organization's records.

A. True
B. False

_____ 3. To ensure complete documentation records should include the proper identification of document originators and document recipients.
A. True
B. False

_____ 4. Records-management systems may be manual or automated.
A. True
B. False

_____ 5. All records-management systems must be in a paper or physical format.
A. True
B. False

_____ 6. Functions of the records-management system include providing security from unauthorized access or tampering.
A. True
B. False

_____ 7. Records provide documentation of required training completion.
A. True
B. False

_____ 8. Records cannot be used in defense of legal challenges in cases of accidents, fatalities, or injuries.
A. True
B. False

_____ 9. Self-study by an individual is difficult to document and should NOT be documented in training records.
A. True
B. False

_____ 10. Training records should be open and no restrictions placed on who can view them.
A. True
B. False

_____ 11. Types of training documented in training records include daily training delivered by a training division.
A. True
B. False

_____ 12. Standardized forms ensure that information that is mandated by law and stored in the records-management system is consistent and complete.

A. True
B. False

_____ 13. Computers can be used to create forms, but it is a difficult and tedious process.

A. True
B. False

_____ 14. Indications that a policy, procedure, or guideline needs to be revised include an increase in policy infractions.

A. True
B. False

_____ 15. The policies, procedures, or guidelines of the training organization need to be evaluated for effectiveness every three to five years.

A. True
B. False

_____ 16. The most common standards used by fire and emergency services organizations in North America are those from NIOSH.

A. True
B. False

_____ 17. Most states and provinces have specific requirements for emergency medical services (EMS) training and certification.

A. True
B. False

_____ 18. When making the transition to a higher rank, the individual should NOT admit mistakes and errors.

A. True
B. False

_____ 19. Selection criteria for instructors should be identical regardless of whether the instructor is short-term or long-term.

A. True
B. False

_____ 20. After instructor roles and qualifications are determined, the training manager must advertise or market the position to prospective candidates both inside and outside the organization.

A. True
B. False

_____ 21. Informal personnel evaluations should NOT be used as a basis for the formal periodic performance review.

A. True
B. False

_____ 22. An advantage of the personnel evaluation program is that it creates a permanent record of an employee's achievements.

A. True
B. False

_____ 23. To ensure a successful formal interview there should be no surprises regarding performance or expectations of the employee during the interview.

A. True
B. False

_____ 24. The job performance of new instructors or staff members does NOT need to be monitored if they are experienced employees.

A. True
B. False

_____ 25. To ensure a successful formal evaluation the employee should be allowed to contribute to the establishing or altering of performance goals and objectives.

A. True
B. False

List

C. Write the types of training documented in training records on the lines provided.

1. ______
2. ______
3. ______
4. ______
5. ______
6. ______

Terms

D. Write the definition of the following terms on the lines provided.

1. Policy ______
2. Procedure ______
3. Guideline ______
4. Adoption ______

Describe

E. **Describe the concept of a 360-degree feedback evaluation on the lines provided.**

Chapter 21 Analysis and Evaluation

Multiple Choice

A. Write the correct letters on the blanks.

_____ 1. Which of the following types of analyses is based on the relationship between the effort and the result?

A. Needs analysis
B. Policy analysis
C. Process analysis
D. Cost/benefit analysis

_____ 2. Which of the following types of analyses involves looking at each step in the process and determining the best way to do it?

A. Needs analysis
B. Policy analysis
C. Process analysis
D. Cost/benefit analysis

_____ 3. Which of the following types of analyses occurs when current policies are analyzed for effectiveness and enforcement?

A. Needs analysis
B. Policy analysis
C. Process analysis
D. Cost/benefit analysis

_____ 4. Which of the following types of analyses is used to determine the most efficient ways to provide a program or service by looking at each program component?

A. Needs analysis
B. Task analysis
C. Risk analysis
D. Program analysis

_____ 5. Which of the following types of analyses is conducted to determine the types of services that an organization is currently delivering and compare them to the services that the community desires?

A. Needs analysis
B. Task analysis
C. Risk analysis
D. Program analysis

_____ 6. Which of the following types of analyses is a systematic appraisal of duties of a specific job or jobs, which identifies and describes all component tasks of that job or jobs?

A. Needs analysis
B. Task analysis
C. Risk analysis
D. Program analysis

_____ 7. Which of the following types of analyses is applied to the likelihood of an event occurring such as a natural disaster?

A. Needs analysis
B. Task analysis
C. Risk analysis
D. Program analysis

_____ 8. Which of the following is a characteristic of prospective or formative evaluations?

A. Relies on feedback
B. Relies on evidence
C. Generates new programs or projects
D. Concerns the overall effectiveness of the program

_____ 9. Which of the following is a characteristic of retrospective or summative evaluations?

A. Provides an ongoing process
B. Provides a look into the future
C. Shapes the direction of professional development
D. Documents habits that were relied upon by students

_____ 10. Which of the following evaluations used for fire and emergency services programs determines how a program actually works and highlights its strengths and weaknesses?

A. Needs-based evaluations
B. Goals-based evaluations
C. Process-based evaluations
D. Outcomes-based evaluations

True / False

B. Write the correct letters on the blanks.

_____ 1. The first step in the analysis process is to determine each of the components that compose the larger item, program, or process.
A. True
B. False

_____ 2. A decision or planning model can be used for selecting the appropriate methods of instruction and budgets to cover costs.
A. True
B. False

_____ 3. In the analysis process, when developing approaches to a problem a minimum of three options is generated: best, compromise, status quo (no action at all).
A. True
B. False

_____ 4. Both the analysis process and the evaluation process are objective.
A. True
B. False

_____ 5. Qualitative evaluation is based on a numeric or statistical analysis.
A. True
B. False

_____ 6. Both internal and external programs of a fire and emergency services organization must be evaluated periodically.
A. True
B. False

_____ 7. Evaluation determines how effective and efficient a person, an item, a program, or a project is compared to a benchmark or established set of criteria.
A. True
B. False

_____ 8. Quantitative evaluation is based most often on nonnumeric analysis.
A. True
B. False

_____ 9. The evaluation of a fire and emergency services organization is an ongoing process.
A. True
B. False

_____ 10. Training program evaluation has three major components: criteria, evidence, and judgment.
A. True
B. False

_____ 11. Areas to address in an evaluation of the instructional process include: reaction, knowledge, skills, attitudes, transfer of learning, and results.
A. True
B. False

_____ 12. A formative evaluation is an end-of-the-course appraisal.
A. True
B. False

_____ 13. The first step to take after reviewing evaluation results is to identify actions to correct deficiencies.
A. True
B. False

_____ 14. Two ways to conduct formative evaluations are observations and field-testing.
A. True
B. False

_____ 15. Two sources of evidence for summative evaluations are test results and course feedback from students.
A. True
B. False

Terms

C. Describe the following terms on the lines provided.

1. Qualitative evaluation ____

____

____

2. Quantitative evaluation ____

____

____

3. Formative evaluation ____

____

____

4. Summative evaluation ____

____

____

5. Field test ____

____

____

Chapter 22 Program and Curriculum Development

Multiple Choice

A. Write the correct letters on the blanks.

_____ 1. Which of the following is the first step in the five-step planning model?
   A. Select
   B. Identify
   C. Design
   D. Implement

_____ 2. In which step of the five-step planning model is a needs analysis performed to determine the training program, curriculum, or course required to meet the organization's needs and jurisdictional mandates?
   A. Select
   B. Identify
   C. Design
   D. Implement

_____ 3. In which step of the five-step planning model are goals established and course outcomes and learning objectives developed?
   A. Select
   B. Identify
   C. Design
   D. Implement

_____ 4. In which step of the five-step planning model is the effectiveness of the course or curriculum determined?
   A. Identify
   B. Design
   C. Implement
   D. Evaluate

_____ 5. In which step of the five-step planning model are lesson plans developed?
   A. Identify
   B. Design
   C. Implement
   D. Evaluate

_____ 6. In which step of the five-step planning model is a pilot presentation of the course or curriculum performed?
A. Identify
B. Design
C. Implement
D. Evaluate

_____ 7. What must be performed when the identification step reveals that change is required or that different or additional services are required?
A. Task analysis
B. Needs analysis
C. Program analysis
D. Policy analysis

_____ 8. Which of the following is a detailed review of each job that is performed by emergency personnel?
A. Task analysis
B. Needs analysis
C. Program analysis
D. Policy analysis

_____ 9. Which of the following terms refers to the grouping of similar functions within a block?
A. Jobs
B. Tasks
C. Task steps
D. Occupations

_____ 10. What part of the lesson plan provides a motivational statement that gets students' attention and interest?
A. Preparation section
B. Presentation section
C. Application section
D. Evaluation section

_____ 11. In what task of the design step must lessons be placed in order so that the most basic knowledge is taught first?
A. Develop lesson plans
B. Group similar objectives
C. Sequence lessons into courses
D. Sequence courses into a curriculum

_____ 12. Which of the following tasks of the design step ensures that objectives match the skills and are in the proper sequence to accomplish the task and therefore the job?
A. Develop lesson plans.
B. Group similar objectives.
C. Sequence lessons into courses.
D. Sequence courses into a curriculum.

_____ 13. Which of the following is typically the first step in the implementation step?
A. Qualify instructors.
B. Select students.
C. Obtain final course approval.
D. Present a pilot version of the course.

_____ 14. Which of the following is typically the last step in the implementation step?
A. Qualify instructors.
B. Select students.
C. Obtain final course approval.
D. Present a pilot version of the course.

_____ 15. Which part of the presentation section provides opportunities for students to practice applying the information or skill?
A. Preparation section
B. Presentation section
C. Application section
D. Evaluation section

True / False

B. Write the correct letters on the blanks.

_____ 1. The identification step of the five-step planning model begins with the realization that a change in the organization's operating environment has occurred.
A. True
B. False

_____ 2. Professional qualification standards can be used for designing and evaluating training and certifying personnel.
A. True
B. False

_____ 3. A goal is a broad objective that states performance required for certification.
A. True
B. False

_____ 4. Learning objectives should contain the following components: conditions description, performance statement, and standards criteria.
A. True
B. False

_____ 5. Performance standards may be dictated by outside entities such as NFPA or legislation.
A. True
B. False

_____ 6. Objectives are used to state general goals but should NOT be the basis for student testing and evaluation.
A. True
B. False

_____ 7. It is NOT necessary to have a certified Level I or Level II instructor to teach courses that are specific to the fire and emergency services.
A. True
B. False

_____ 8. In the implementation step the instructor should create or select appropriate training aids.
A. True
B. False

_____ 9. Evaluating the instructor is NOT considered a part of the evaluation step in the five-step planning model.
A. True
B. False

_____ 10. A summative evaluation only needs to be performed after the initial course presentation.
A. True
B. False

_____ 11. During the presentation of a course, a course instructor should be periodically evaluated by more experienced instructors with subject-matter expertise.
A. True
B. False

_____ 12. When planning curriculum/course revisions involve other branches or divisions after any proposed changes have already been implemented.

A. True
B. False

_____ 13. When planning curriculum/course revisions create clear, concise revision proposals that meet the criteria for any new curriculum or course.

A. True
B. False

_____ 14. In the evaluation step of the five-step planning model there must be an evaluation system for course materials and instructor performance.

A. True
B. False

_____ 15. Low test scores always indicate that a student did NOT learn.

A. True
B. False

Describe

C. Briefly describe the steps in the five-step planning model on the lines provided.

1. ____

____

____

____

____

2. ____

____

____

____

____

3. ______________________________

____

____

____

____

4. ______________________________

____

____

____

____

5. ______________________________

____

____

____

____

Chapter 1 Answers

Multiple Choice

A.

1. C (24)
2. C (25)
3. B (24)
4. A (10)
5. C (13)
6. C (16)
7. B (20)
8. A (17)
9. D (14)
10. B (29)
11. D (30)
12. D (28)
13. D (29)
14. A (21)
15. C (20)

True/False

B.

1. A (13)
2. B Instructors should not attempt to bluff their way through quickly because that can cost them credibility with students. (21)
3. A (22)
4. B Instructors should not expect all students to act, think, respond, or learn in the same way or at the same rate. (23)
5. B Instructors should not depend on one source for information, especially a product vendor or manufacturer. (27)
6. A (29)
7. A (30)
8. B Always use the adopted accountability system and the Incident Command System at all training sessions. (28)
9. A (28)
10. A (25)
11. B An instructor who attempts to use intimidation to get results will not stimulate students to reach their full potentials. (22)
12. A (22)
13. A (11)
14. A (11)
15. A (13)

List

C. Effective instructor characteristics should include the following items in any order:

1. Leadership abilities (16)
2. Strong interpersonal skills (16)
3. Subject and teaching competencies (16)
4. Desire to teach (16)
5. Enthusiasm (16)
6. Motivation (16)
7. Preparation and organization (16)
8. Ingenuity, creativity, and flexibility (16)
9. Empathy (16)
10. Conflict-resolution skills (16)
11. Fairness (16)
12. Personal integrity (16)

Chapter 2 Answers

Multiple Choice

A.

1. B (40)
2. A (42)
3. A (41)
4. B (44)
5. B (45)
6. B (47)
7. C (47)
8. D (52)
9. C (53)
10. B (55)

True/False

B.

1. A (35)
2. B The instructor's responsibility is to balance realism and safety and ensure a safe training environment. (40-41)
3. A (35)
4. B One recommendation for casualty prevention is to include the use of live-burn evolutions in a variety of structure types. (37)
5. A (40)
6. A (42)
7. B The presence of an incident safety officer does not relieve instructors from monitoring training and emphasizing safety. (43)
8. B The health and safety officer is responsible for teaching safety-related topics but that responsibility may become the duty of an Instructor I under certain conditions. (44)
9. A (44-45)
10. B When conducting practical training evolutions, instructors must either act as ISO or appoint a safety officer who acts as ISO or monitor. (46)
11. B Accidents are usually the result of unsafe acts by persons who are unaware or uninformed of potential hazards, are ignorant of the safety policies, or fail to follow safety procedures. (47)
12. A (49)
13. A (50)
14. B The ISO is generally responsible for the accident investigation as required in NFPA 1521. (50)
15. A (47)

List

C. Places where injuries and fatalities can occur should include the following items in any order:

1. When responding to and from incidents (35)
2. At the incident scene (35)
3. During training (35)
4. During work shifts at the station (35)

Terms

D.

1. Task analysis. A task analysis consists of a detailed review of each physical task or job that is performed by emergency personnel. (41)
2. Hazard/risk analysis. A hazard/risk analysis identifies potential problem areas and is the foundation for a risk-management plan. (41)
3. Incident Action Plan. The IAP establishes the strategic goals and tactical objectives of the operation or scenario for a specific time period. (42)

Chapter 3 Answers

Multiple Choice

A.

1. B (63)
2. D (62)
3. A (62)
4. A (67)
5. C (67)
6. B (64)
7. A (64)
8. D (65)
9. C (64)
10. C (66)
11. D (69)
12. B (71)
13. C (72)
14. C (73)
15. D (74)
16. B (76)
17. D (76)
18. C (77)
19. D (77)
20. B (78)

True/False

B.

1. A (61)
2. A (62)
3. B The most commonly known standards in the fire and emergency services are the consensus standards developed by the National Fire Protection Association. (63)
4. B The three types of law are legislative, administrative, and judiciary. (64)
5. A (64)
6. A (66)
7. B An ordinance is a local law that applies to persons, things, and activities in a jurisdiction and has the same force and effect as statutory law. (66)
8. A (68)
9. B The length of time that records must be retained by the organization depends on state/provincial and local laws and the specific type of record. (69)
10. B Official meeting minutes and any other notes that are made as part of a meeting are part of public records. (69)
11. A (69)
12. A (70)
13. B Exposing students to hazardous training environments without having adequate controlling elements could be perceived as negligence in court if an injury occurs. (70)
14. B Instructors are expected to foresee (predict) potential injury events and prevent injuries while training personnel. (70)
15. A (72)
16. A (73)
17. B Individuals have the right to control the use of pictures of themselves and their property. (74)
18. A (75)
19. A (79)
20. B When resolving an ethical dilemma, the individual should conduct an objective investigation to gather the details of the event. (80)

List

C. Ways that instructors can reduce the potential for liability and legal action against themselves and their organizations should include the following items in any order:

1. Be aware of standard expectations (70)
2. Teach to the standards (70)
3. Teach only topics they are qualified to teach (70)
4. Provide a safe learning environment (70)

Describe

D.

1. Fair use doctrine of the Copyright Act. The fair use doctrine of the Copyright Act grants the privilege of copying materials to persons other than the owner of the copyright without consent when the material is used in a reasonable manner. (73)
2. Invasion of privacy. Invasion of privacy is the wrongful intrusion into a person's private activities by the government or other individuals. (74)

Chapter 4 Answers

Multiple Choice

A.

1. B (87)
2. D (86)
3. D (88)
4. A (88)
5. C (89)
6. D (96)
7. B (95)
8. B (96)
9. D (87)
10. C (91)
11. D (91)
12. D (92)
13. B (91)
14. D (93)
15. D (95)

True/False

B.

1. A (87)
2. B The preferred approach to radio communication in the fire service is clear-text (plain English). (87)
3. B Interference is NOT an essential element of interpersonal communication. (86)
4. A (90)
5. B Instructors should avoid the use of technical language and fire service jargon when speaking with individuals from outside the profession. (90)
6. A (91)
7. A (95)
8. A (96)
9. B Taking notes during formal speeches is a good way to help improve listening skills. (96)
10. A (96)

List

C. The essential basic elements of communication should include the following items in any order:

1. Sender (86)
2. Message (86)
3. Medium or channel (86)
4. Receiver (86)
5. Feedback to the sender (86)

Describe

D.

1. Learning — Acquire knowledge or skills (89)
2. Relating — Establish a new relationship or maintain an existing one (89)
3. Influencing — Control, direct, or manipulate behavior (89)
4. Playing — Create a diversion and gain pleasure or gratification (89)
5. Helping — Minister to the needs of another person or console someone in the time of tragedy or loss (89)

Chapter 5 Answers

Multiple Choice

A.

1. B (101)
2. C (102)
3. C (102)
4. A (105)
5. C (106)
6. C (109)
7. B (108)
8. C (109)
9. C (112)
10. C (114)

True/False

B.

1. A (103)
2. B Training areas need to be secured from public access to prevent vandalism and injury to the public. (104)
3. B Driving courses may include parked vehicles, overhead wires, or other obstacles to simulate actual driving situations. (105)
4. A (107)
5. A (107-108)
6. A (109)
7. B Although only a simulated hazardous material is involved in the training and training foam concentrate is used in the exercises, runoff water must still be contained and decontaminated. (110)
8. A (111)
9. A (112)
10. B Acquired structures used for live-fire training must meet the safety requirements of NFPA 1403. (113)
11. A (113)
12. A (114)
13. B When environmental conditions are detrimental to learning or are inherently unsafe, instructors may have to limit the training exercises or not allow the training to occur. (113)
14. B Water rescue training can be performed in local ponds or lakes, rivers and streams, protected coastal areas, and swimming pools. (106)
15. A (115)

List

C. Live-fire training area characteristics should include the following items in any order:

1. Constructed to meet the guidelines set forth in NFPA 1402 and used to fulfill the requirements of NFPA 1403 (111)
2. Located remotely from other facilities or occupancies that may be affected by the smoke and heat of a fire (111)
3. Located to protect the soil from contaminated runoff water (111)
4. Equipped with thermal sensing devices to determine actual temperature within the structure (111)
5. Equipped with a fuel shutoff or automatic fire-suppression system in the event of an accident (111)

Chapter 6 Answers

Multiple Choice

A.

1. C (121)
2. B (121)
3. D (121)
4. C (123)
5. D (123)
6. D (127)
7. D (127)
8. C (128)
9. D (128)
10. C (130)
11. A (130)
12. C (128)
13. D (127)
14. A (123)
15. A (121)

True/False

B.

1. B The majority of reports that instructors write are public records and may be read by people outside the organization. (120)
2. A (121)
3. A (119)
4. B An executive summary is a brief review of the key points in a report. (123)
5. A (120)
6. A (128)
7. B Preventive maintenance is performed to prevent damage from occurring and extend the useful life of an item, vehicle, or facility by reducing wear. (129)
8. A (123)
9. A (126)
10. B Information stored in a record-keeping system can be used to justify budget requests or program development. (125)
11. A (125)
12. A (126)
13. B Keeping accurate training records identifies training areas that are emphasized as well as areas that require more attention. (127)
14. A (127)
15. A (127)
16. A (130)
17. B The first step in determining the record-keeping requirements of an organization is to identify research requirements. (130)
18. A (131)
19. B When implementing a new record-keeping system, some files and documents that already exist will have to be entered into the system or assigned to a hardcopy file or archive. (131)
20. B Annual checks of a record-keeping system can be made following the first year to ensure its effectiveness. (133)

List

C.

The four parts of a report should include the following items:

1. Heading (121)
2. Introduction (121)
3. Body (121)
4. Conclusion/summary (121)

Terms

D.

1. Preventive maintenance. Preventive maintenance is performed to prevent damage from occurring and extend the useful life of an item, vehicle, or facility by reducing wear. (129)
2. Corrective maintenance. Corrective maintenance is performed to repair an item because of an accident, overuse, operator error, or even abuse. (129)
3. Data. Data are the raw material from which information is derived. (133)
4. Analysis. An analysis looks at the relationship between the key elements of the data and between similar information that is gathered from other incident reports. (133)

Chapter 7 Answers

Multiple Choice

A.

1. B (140)
2. B (140)
3. D (140)
4. C (140)
5. B (141)
6. B (143)
7. C (143)
8. A (143)
9. B (143)
10. D (143)
11. B (146)
12. A (146)
13. C (146)
14. B (146)
15. D (147)
16. A (148)
17. D (149)
18. C (156)
19. B (156)
20. D (157)

True/False

B.

1. B In the synthesis level of learning in the cognitive domain students put parts together to form a new whole and they integrate parts to invent new procedures. (143)
2. A (138)
3. B The least understood domain of learning is the affective domain. (144)
4. B Pedagogy is the principle of learning most often associated with children. (137)
5. A (145)
6. A (141)
7. B The cone of learning model states that individuals retain approximately 90 percent of what they say while doing what they are talking about. (150)
8. A (144)
9. A (148)
10. B Long-term memory is considered permanent storage. (153)
11. A (156)
12. A (148)
13. B External motivations such as rewards, recognition, and certificates can increase the adult student's motivation to learn. (149)
14. A (153)
15. B One method of motivating students is to promote working together in peer groups to share and learn other methods. (148)
16. A (153)
17. B Learning plateaus are normal and instructors should work with students to help them overcome problems. (154)
18. B The mastery approach to teaching works very well in the fire and emergency services because the services are competency-based. (157)
19. A (157)
20. A (157)
21. B An advantage of the mastery approach to teaching is that instructors must state the learning objectives before designating or designing student activities and projects. (157)
22. A (157)
23. B Norm-referenced assessments measure the accomplishments of one student against that of another. (158)
24. B Norm-referenced assessments are rarely used in the fire and emergency services. (158)
25. A (158)

List

C. The levels of learning in the cognitive domain from simple to complex are as follows:

1. Knowledge (143)
2. Comprehension (143)
3. Application (143)
4. Analysis (143)
5. Synthesis (143)
6. Evaluation (143)

D. The levels of learning in the psychomotor domain from simple to complex are as follows:

1. Observation (143)
2. Imitation (143)
3. Adaptation (143)
4. Performance (143)
5. Perfection (143)

Describe

E.

1. Cognitive domain. Through the cognitive domain, students gain understanding about a concept or topic. (141)
2. Psychomotor domain. Through the psychomotor domain, students perform the skills associated with that concept or topic. (141)
3. Affective domain. Through the affective domain, students develop a willingness to perform the behavior correctly and safely. (141)

Chapter 8 Answers

Multiple Choice

A.

1. A (167)
2. D (167)
3. A (167)
4. C (171)
5. B (171)
6. A (178)
7. C (182)
8. D (183)
9. B (184)
10. A (184)
11. C (167)
12. D (170)
13. A (174)
14. B (174)
15. A (184)

True/False

B.

1. B Instructors should use discussion techniques to establish a relationship between past experiences and new information. (164)
2. A (165)
3. A (169)
4. B Literacy level refers to reading and comprehension ability. (169)
5. A (172)
6. B Instructors should avoid calling on shy or timid individuals for discussion or response until the student is comfortable in the class. (173)
7. A (178)
8. A (175)
9. B Behavioral management is one of the most important skills for an instructor to acquire and maintain. (175)
10. A (177)
11. B Instructors must NOT assume the role of therapist when a student appears to have a psychological/ emotional problem. (177)
12. A (178)
13. A (182)
14. A (177)
15. A (185)

List

C. Student-originated disruptive behavior should include the following items in any order:

1. Arriving late (183)
2. Speaking out loud (183)
3. Talking with others off the subject (183)
4. Sleeping in class accompanied by snoring (183)
5. Showing off intentionally (183)
6. Interrupting others (183)
7. Sidetracking discussions (183)
8. Seeking attention (183)
9. Acting blatantly insubordinate and disrespectful (183)

Describe

D.

1. Learning disabilities. Learning disabilities consist of a wide variety of disorders that may be neurological in origin and affect an individual's ability to understand, think, use the spoken or written word, perform mathematical functions, or perform fine psychomotor skills. (171)
2. Coaching. Coaching is the process of giving motivational correction, positive reinforcement, and constructive feedback to students in order to maintain and improve their performance. (178)
3. Mentoring. Mentoring places a new student under the guidance of a more experienced professional or another student who acts as tutor, guide, and motivator. (182)
4. L-E-A-S-T method of progressive discipline.
   Leave it alone—Notice whether the behavior goes away.
   Eye contact—Look at the student long enough to make eye contact.
   Action—Take action when the behavior continues.
   Stop the class—Stop the class and discuss the problem with the student.
   Terminate the student—Expel the student from the class. (185)

Chapter 9 Answers

Multiple Choice

A.

1. D (196)
2. D (190)
3. C (191)
4. D (193)
5. C (195)
6. A (193)
7. B (195)
8. C (207)
9. A (207)
10. C (209)
11. D (209)
12. C (195)
13. B (195)
14. A (193)
15. B (190)

True/False

B.

1. B An instructor who is unqualified to teach a topic should NOT attempt to teach it. (198)
2. A (195)
3. A (197)
4. B The primary role of the Level I Instructor is to teach from a prepared lesson plan. (197)
5. A (190)
6. A (196)
7. A (199)
8. A (201)
9. A (202)
10. B When scheduling equipment, an important factor in maintaining the continuity of the course is using the same type of equipment in the learning sessions that is used in the testing session and on the job. (203)
11. A (204)
12. B Consistency in application of safety procedures ensures that students will continue to follow the procedures during emergency incidents. (205)
13. B Frequent rest breaks, usually after every 45 to 50 minutes of instruction, allow students to stand, move around, stretch, and attend to other comfort needs. (207)
14. A (207)
15. B The primary source of light for an indoor classroom should be fluorescent lights. (209)
16. A (211)
17. B It is the instructor's responsibility to eliminate safety hazards, such as tripping hazards and electrical cords, in the indoor classroom. (212)
18. B Outdoor training sessions can occur along public streets or in parking lots that may be affected by vehicle traffic. (213)
19. A (215)
20. B Realistic training scenarios are just as noisy as actual emergency incidents. (213)

List

C. The four steps in the four-step method of instruction include the following:

1. Preparation (193)
2. Presentation (193)
3. Application (193)
4. Evaluation (193)

D. Considerations should include the following in any order:

1. Weather conditions (213)
2. Terrain (213)
3. Vehicle traffic (213)
4. Vehicle and machine noise (213)
5. Light levels (215)
6. Site space (215)
7. Exposures (215)
8. Environmental laws and codes (215)
9. Access (215)

Describe

E.

1. Job or topic—Short descriptive title of the information covered (190)
2. Time frame—Estimated time it takes to teach the lesson (190)
3. Level of instruction—Desired learning level that students will reach by the end of the lesson (190)
4. Learning objectives—Descriptions of the minimum acceptable behaviors that students must display by the end of an instructional period (190)
5. Resources/materials needed—List of all items needed to teach the number of students in the course (190)
6. Prerequisites—List of information, skills, or previous requirements that students must have completed or mastered before entering this course or starting this lesson (191)
7. References—List of specific references and resources on the lesson plan along with page numbers to refer to and review (193)
8. Lesson summary—Restatement or reemphasis of the key points of the lesson to clarify uncertainties, prevent misconceptions, increase learning, and improve retention (193)
9. Assignments—Readings, practice, research, or other outside-of-class requirements for students (193)
10. Lesson outline—Summary of the information to be taught (193)
11. Evaluations—Type of evaluation instrument the instructor will use to determine whether students have met lesson objectives (193)

Chapter 10 Answers

Multiple Choice

A.

1. D (221)
2. C (222)
3. D (224)
4. C (227)
5. D (230)
6. D (235)
7. A (236)
8. A (237)
9. B (237)
10. C (237)
11. C (237)
12. D (239)
13. D (240)
14. B (244)
15. D (244)

True/False

B.

1. B A disadvantage of the lecture format is that there are limited senses involved in receiving the information. (221)
2. B Students must have a basic knowledge of the subject before the discussion begins. (223)
3. A (227-228)
4. A (221)
5. B The discussion presentation format works best for small groups of approximately 3 to 15 students. (223)
6. A (223)
7. A (225)
8. B When demonstrating a skill, allow students the opportunity to ask questions and clarify any misunderstandings. (225)
9. A (225)
10. B The multiple instructor presentation format requires more hours per instructor to accomplish the same course requirements and requires more advanced planning than individual teaching. (226)
11. B An advantage of the multiple-instructor presentation format is that it works well when the topic is broad. (226)
12. A (226)
13. A (228)
14. A (228)
15. B Security is a major issue of concern with technology-based training. (228)
16. B In self-directed learning an instructor is NOT directly involved in the delivery of the training. (230)
17. B Not all types of training programs, particularly basic-level skill programs, are suited to self-directed learning. (231)
18. A (230)
19. A (232-233)
20. A (236)
21. A (238)
22. B Repetition is an integral part of active learning and the organization of the presentation. (241)
23. A (243)
24. B Guidelines for asking effective questions include asking only one question at a time and never using questions to embarrass students. (243)
25. B Never bluff students by providing false or misleading information. (245)

Terms

C.

1. Instructor-led training—Traditional instruction that depends on the direct transfer of knowledge from the instructor to the student (220)
2. Technology-based training—Electronic learning that uses methods such as Internet web-based instruction, interactive television (ITV), and other forms of computer-based electronically transferred knowledge (220)
3. Blended electronic learning—Learning that combines online learning courses that students complete independently with classroom-delivered instruction and hands-on, performance-based skills instruction (228)
4. Individualized instruction—The process of matching instructional methods with learning objectives and individual learning styles (231)
5. Active learning—Using instructional activities that involve students in doing things and thinking about what they are doing (238)

Chapter 11 Answers

Multiple Choice

A.

1. A (250)
2. C (254)
3. C (254)
4. D (255)
5. A (256)
6. C (258)
7. C (259)
8. A (262)
9. D (263)
10. A (264)
11. B (267)
12. D (266)
13. C (271)
14. D (270)
15. D (267)

True/False

B.

1. A (251)
2. B Standardized training curricula has made it easier for instructors to select appropriate audiovisual training aids. (251)
3. A (257)
4. A (254)
5. B A guideline for media transitions is to keep backgrounds simple so that they do not conflict with the text or graphics. (258)
6. A (256)
7. A (258)
8. B Rear-screen projection systems are more expensive than front-screen projection devices. (265)
9. A (258)
10. B Level I Instructors should be familiar with equipment used in the completion of their duties. (271)
11. A (271)
12. B Electrical power cords need to be unplugged before opening audiovisual equipment. (274)
13. B Air filters in multimedia projectors need to be cleaned periodically. (274)
14. A (273)
15. A (273)
16. A (273)
17. B Care guidelines for audiovisual training aids and devices include placing dust covers over equipment when not in use. (273)
18. A (273)
19. A (273)
20. B Cleaning guidelines for audiovisual training aid devices include NOT using solvents for cleaning training aids. (273)

List

C. Purposes of transitions should include the following items in any order:

1. Maintain interest (255)
2. Maintain continuity (255)
3. Maintain consistency (255)
4. Establish relationships (255)
5. Provide previews (255)
6. Provide summaries (255)

Chapter 12 Answers

Multiple Choice

A.

1. B (277)
2. C (279)
3. C (280)
4. C (279)
5. A (281)
6. A (281)
7. A (282)
8. D (283)
9. B (290)
10. C (290)
11. D (295)
12. B (295)
13. D (295)
14. C (296)
15. A (296)
16. D (296)
17. C (297)
18. A (296)
19. C (297)
20. D (297)

True/False

B.

1. B For experienced firefighters, live-fire and other technical training evolutions provide opportunities to develop additional skills and increase skill-performance levels. (283)
2. A (283)
3. B Training evolutions may be held at either permanent training facilities or remote sites. (284)
4. A (283)
5. B Instructors must balance the level of realism in the training evolution with the level of risk to the safety and health of students. (285)
6. A (285)
7. B The National Incident Management System (NIMS) ICS must be established and followed whether the evolution is intended to involve a single company, multiple companies, or multiple agencies and jurisdictions. (286)
8. B The instructor-to-student ratio depends on the specific evolution. (288)
9. A (288)
10. A (289)
11. B While monitoring the practical training evolution, the instructor should immediately stop and correct any performance weaknesses or errors. (290)
12. A (290)
13. A (291)
14. A (292)
15. B Live-fire training is an essential part of firefighter training. (292)
16. A (293)
17. B The majority of fires involve small quantities of fuel. (293)
18. B During interior structural fire training, students or instructors may NOT act as victims within the burn structure. (294)
19. B During interior structural fire training, students must wear full PPE. (294)
20. A (297)
21. B During agricultural equipment rescue operations training, mannequins are used to simulate victims. (296)
22. A (295)
23. A (298)
24. B Instructors should receive a refresher course before instructing students or supervising practical training evolutions. (298)
25. A (298)

List

C. Factors to consider when planning practical evolutions should include the following items in any order:

1. Safety (285)
2. Learning objective/outcome (285)
3. Justification (285)
4. Supervision (286)
5. Resources/logistics (286)
6. Weather (286)
7. Legal requirements (286)
8. Incident Command System (286)
9. Coordination (286)
10. Exposures (286)
11. Evaluation/critique (287)

Describe

D.

1. Case study—A description of a real or hypothetical problem that an organization or an individual has dealt with or could deal with (277)
2. Role-plays—In role-plays, students portray characters and act the roles assigned to them in scenarios. (279)
3. Brainstorming sessions—In the brainstorming method, a group of students is given a problem or situation and time to determine a solution to it. (280)
4. Simulations—Training simulations are activities that allow students to participate in scenarios that represent real-life situations. (281)
5. Field and laboratory experiences—Field and laboratory experiences involve elements of the demonstration and simulation methods where students have the opportunity to inspect, use, test, and evaluate equipment or processes either in actual installations or laboratory settings. (282)

Chapter 13 Answers

Multiple Choice

A.

1. A (303)
2. B (304)
3. B (305)
4. A (305)
5. C (305)
6. C (307)
7. B (307)
8. D (307)
9. C

True/False

B.

1. B Norm-referenced tests rate student performance compared to other students based on broad sampling. (305)
2. B Oral tests are NOT commonly used in the fire and emergency services. (306)
3. A (303)
4. A (304)
5. A (305)
6. B Oral tests should never be used as the sole means of evaluating students for terminal performance or officer candidates for promotion. (307)
7. A (306)
8. B Written tests may be subjective or objective. (307)
9. B During performance tests, students are tested on their present abilities. (308)
10. B Cheating on tests presents special problems for instructors. (314)
11. A (308)
12. B Scoring performance tests can be very subjective. (311)
13. B When scoring written tests, add comments to essay or short-answer questions. (311)
14. A (311)
15. A (310)
16. A (314)
17. B Students who fail to achieve the minimum required grade should NOT be certified as having completed the training. (314)
18. A (315)
19. A (316)
20. B Feedback that is one-sided with instructors providing the criticism and the solution can have negative results. (316)
21. A (316)
22. A (316)
23. B In order to maintain test security, do NOT rely on questions published either in a textbook or study guide. (315)
24. B In order to maintain test security, regularly revise test questions and answer sheets. (314)
25. A (307)

Terms

C.

1. Criterion-referenced tests compare performance against appropriate minimum standards. (305)
2. Norm-referenced tests rate student performance compared to other students based on a broad sampling. (305)
3. Prescriptive tests are given at the beginning of instruction to establish a student's current level of knowledge in order to measure readiness and placement. (305)
4. Formative tests are often viewed as quizzes, pop tests, or question/answer periods in class that are given throughout the course or unit of instruction to measure improvement. (305)
5. Summative tests measure student achievement in an entire area on a number of topics covered over a long period of time. (305)

Chapter 14 Answers

Multiple Choice

A.

1. D (323)
2. C (323)
3. B (328)
4. A (328)
5. D (322-323)
6. B (330)
7. D (331)
8. B (341)
9. D (346)
10. C (339)
11. A (346)
12. A (349)
13. B (351)
14. A (350)
15. C (351)

True/False

B.

1. B Learning objectives should be specific and objective. (331)
2. B One guideline for developing effective visual aids is to use a maximum of 6 lines down and 6 words across the viewing area. (349)
3. A (322)
4. A (322)
5. B The learning objective model developed by Mager consists of three components: the conditions description, the performance statement, and the standards criteria. (330)
6. A (329)
7. A (329)
8. B A variety of formats can be used to write a lesson plan. (333)
9. A (330)
10. B The first step in creating a lesson plan is to identify the expected performance outcomes. (338)
11. B When the majority of students did not perform satisfactorily, the lesson plan may need to be revised. (352-353)
12. B When it is determined that there is a problem with a lesson plan, the instructor and/or training division should revise it. (353)
13. A (338)
14. A (333)
15. A (339)

Describe

C.

1. Conditions description—Describe the situation, tools, or materials required for a student to perform a single specific action or behavior (330)
2. Performance (behavior) statement—Identify what the student is expected to do (330)
3. Standards criteria—State the acceptable level of student performance (330)

Chapter 15 Answers

Multiple Choice

A.

1. A (363)
2. C (358-359)
3. A (364)
4. B (364)
5. A (359)
6. D (360)
7. C

True/False

B.

1. B The strengths and weaknesses of an employee that are discussed during a personnel evaluation become part of a permanent record. (358)
2. A (358)
3. B When critiquing instructor performance either informally or formally, the instructor-supervisor should first provide positive comments. (359)
4. B When conducting formal instructor evaluations, the instructor supervisor should discuss the evaluation process with the instructor being evaluated. (359)
5. A (359)
6. B The subordinate being evaluated should be included in the decision-making process used to apply the strengths constructively or correct the weaknesses identified in an evaluation. (361)
7. A (361)
8. A (361)
9. B Never send a survey home with students and expect them to return it. (362)
10. B Students should NOT be permitted to sign their names on evaluations. (362)
11. A (362)
12. A (363)
13. A (363)
14. A (365)
15. B When using observation as a method for conducting formative evaluations, the evaluation occurs throughout instruction. (364)

Describe

C.

1. Reaction—Were students satisfied with the course? If not, why? Were instructors and management officials satisfied with the learning that occurred? If not, why? (363)
2. Knowledge—What new knowledge did students acquire and demonstrate? (363)
3. Skills—What new skills did students acquire and demonstrate? (363)
4. Attitudes— How has the training changed students' opinions, values, and beliefs? (363)
5. Transfer of learning—How has the training affected the ways students perform on their jobs? (363)
6. Results—How has the training contributed to accomplishing organizational goals and objectives? (363)

Terms

D.

1. Formative evaluation—The ongoing, repeated checking during course development and during instruction to determine the most effective instructional content, presentation methods, training aids, and testing techniques (363)
2. Summative evaluation—An end-of-the-course appraisal that commonly measures learning by some form of objective or subjective evaluation instrument (364)
3. Field test—The process of teaching the course on a trail basis (364)

Chapter 16 Answers

Multiple Choice

A.

1. C (377)
2. D (377)
3. D (376)
4. D (377)
5. C (377)
6. D (377)
7. B (377)
8. A (377)
9. D (380)
10. D (382)
11. B (383)
12. B (383)
13. C (384)
14. B

True/False

B.

1. A (376)
2. A (377)
3. B The first step in test planning is to determine test purpose and type. (376)
4. A (378)
5. B Learning objectives are generally created in a hierarchal format with the course or learning outcomes at the top stated in very general terms. (377)
6. A (384)
7. A (384)
8. B Validity is the extent to which a test measures what it is supposed to measure. (386)
9. A (387)
10. A (384-385)
11. B Reliability is the consistency and accuracy of test measurements. (386)
12. A (391)
13. B A subjective test item has no single correct answer. (390)
14. B When creating multiple-choice test items, place the correct answers in varied positions among the A, B, C, and D choices. (391)
15. B When creating true-false test items, distribute an equal number of true and false items randomly throughout the test. (394)
16. A (391)
17. A (394)
18. A (396)
19. B When creating matching test items arrange responses in a systematic manner. (397)
20. A (397)
21. A (397)
22. B A guideline for creating short-answer/completion test items is to arrange the statement in order to place the blanks at or near the end of the sentence. (398)
23. A (399)
24. B A disadvantage of short-answer/completion test items is that they may be difficult to score. (399)
25. A (399)
26. A (401)
27. B An advantage of essay test items is that they are easy to create and can test higher learning levels (399)
28. A (401)
29. B Oral tests are NOT commonly used in the fire and emergency services. (401)
30. B When the purpose of an oral test is to determine knowledge, then the questions should be closed. (401-402)
31. A (402)
32. A (405)
33. A (405)
34. B Grading is the act of assigning a value to the score. (406)
35. B Scoring is the act of identifying which answers are right and which are wrong. (406)

36. B Using the criterion-referenced grading system can result in the majority of students either passing or failing the lesson or course. (406)
37. A (407)
38. B Instructors must be prepared to defend any score or grade given. (407)
39. A (411)
40. A (411)

List

C.

1. Step 1: Determine test purpose and type. (376)
2. Step 2: Identify and define learning objectives or learning outcomes. (376)
3. Step 3: Prepare test specifications. (376)
4. Step 4: Construct appropriate test items. (376)

D. Answer should include the following in any order:

1. Written tests (387)
2. Oral tests (387)
3. Performance tests (387)

Describe

E. Answer should include points from the following:
An interpretive exercise consists of introductory material that may be a paragraph of text describing a situation or scenario, numerical data, illustration, graph, table, chart, diagram, or map, followed by a series of test items. Test items may be multiple-choice, true-false, short-answer/completion, matching, or essay. Students read the texts or look at the illustrations and then answer the questions posed in the test items. (401)

Chapter 17 Answers

Multiple Choice

A.

1. B (418)
2. B (418)
3. C (420)
4. A (420)
5. B (420)
6. D (420)
7. C (424)
8. A (424)
9. C (424)
10. B (424)
11. D (424)
12. B (426)
13. A (425)
14. D (426)
15. B (427)

True/False

B.

1. B Announce discussions in advance and provide the topic to students to allow time for them to research and develop their ideas and opinions. (418)
2. B When an instructor decides that a discussion is appropriate for a class, it is important to begin slowly. (417-418)
3. A (417)
4. B An advantage of discussions is that they build interdependence between students and encourage teamwork. (418)
5. A (419)
6. B In small group discussions, the instructor is NOT part of the group. (419)
7. B In large group discussions, the instructor selects the topic. (419)
8. A (420)
9. A (421)
10. B When leading a discussion, the role of the gatekeeper is to ensure that all students have an opportunity to speak and no one dominates the discussion. (421)
11. B The Level II instructor has the responsibility of being familiar with and implementing the procedures of the ICS adopted by the local jurisdiction. (421)
12. B ICS is officially a part of the National Incident Management System (NIMS) and its use is mandatory by federal agencies that receive federal funding. (422)
13. A (422)
14. A (424)
15. A (426)

List

C. Incident command system (ICS) components should include the following items in any order:

1. Common terminology (424)
2. Modular organization (424)
3. Integrated communications (424)
4. Unified command structure (424)
5. Consolidated action plans (424)
6. Manageable span of control (424)
7. Predesignated incident facilities (424)
8. Comprehensive resource management (424)

Terms

D.

1. IAP—Incident Action Plan (424)
2. ICS—Incident Command System (421)
3. NIMS—National Incident Management System (422)
4. ISO—Incident Safety Officer (424)

Chapter 18 Answers

Multiple Choice

A.

1. D (432)
2. C (432)
3. C (434)
4. D (434)
5. D (438)
6. A (439)
7. D (440)
8. B (441)
9. B (441)
10. A (441)
11. C (440)
12. B (443)
13. C (444)
14. A (446)
15. B (446)
16. B (448)
17. C (448)
18. C (454)
19. C (456)
20. B (455)

True/False

B.

1. B Information obtained from vendors/manufacturers of products should be compared to other information. (438)
2. A (436)
3. A (433)
4. B Instructors should use citations for information that is not original. (439)
5. B In the analysis process the best approach is to develop at least two approaches to the problem. (442)
6. A (436)
7. B Developing and managing budgets are highly specialized skills that are usually performed by financial officers who work for the jurisdiction. (442)
8. A (443)
9. B Funds donated through grants/gifts for capital purchases must be used for that purpose only. (445)
10. A (444)
11. A (445)
12. B An effective purchasing process must be an objective process based on fact. (450)
13. B When conducting a review of product data for the purchasing process, units that do NOT meet criteria should be eliminated from consideration. (455)
14. A (450)
15. A (455)

List

C. Sources of information should include the following in any order:

1. Internet (433)
2. Government agencies (433)
3. Libraries (433)
4. Educational institutions (433)
5. Professional organizations (433)
6. Testing and standards organizations (433)
7. Vendors/manufacturers (433)
8. Nonprofit organizations (433)

Terms

D. Sources of information should include the following in any order:

1. Search engines—Vehicles used to locate information on the World Wide Web (434)
2. Capital budget—Budget that includes projected major purchases—items that cost more than a certain specified amount of money and are expected to last more than 1 year, usually 3 years or more (444)
3. Operating budget—Budget that is used to pay for the recurring expenses of the day-to-day operation of the fire and emergency services organization (444)

Chapter 19 Answers

Multiple Choice

A.

1. A (470)
2. D (463)
3. D (470)
4. D (467)
5. C (470)
6. C (469)
7. B (470)
8. A (472)
9. C (473)
10. A (472)

True/False

B.

1. B When the first sign of a problem arises, such as excessive griping, the supervisor should respond in a proactive manner. (463)
2. A (463)
3. A (463)
4. B In order to create an effective team, ensure that teams are composed of people from differing backgrounds. (466)
5. A (464)
6. B In order to create an effective team, the supervisor should allow team members to determine whether the team and team members are successful. (466)
7. A (466-467)
8. B When completing tasks, final responsibility and authority must always remain with the supervising instructor. (467)
9. B As much responsibility and authority as possible must be delegated to the employees, which give them a sense of ownership in the project. (467)
10. A (468)
11. B An employee in the fire and emergency services rarely performs the same task all the time. (468)
12. A (469)
13. A (469-470)
14. A (471)
15. B The Level II instructor has the responsibility of scheduling resources and instructional delivery. (471)
16. A (472)
17. B Student availability is a factor when scheduling training sessions. (473)
18. A (471-472)
19. B Governmental mandates have a tendency to change over time. (473)
20. A (472)

List

C. Ways to create job interest should include the following items in any order:

1. Empower employees (466)
2. Reward employees (466)
3. Celebrate accomplishments (467)

D. Nine major responsibilities should include the following items in any order:

1. Set a clear and positive example for subordinates. (467)
2. Receive assignments and complete tasks or objectives efficiently and effectively. (467)
3. Promote and maintain health and safety policies within the workplace. (467)
4. Develop an environment of cooperation and teamwork. (467)
5. Promote skills development, skills maintenance, and skills improvement in employees. (467)
6. Maintain discipline. (467)
7. Promote the pursuit of educational and professional opportunities. (467)
8. Promote credentialing and certification as opportunities to enhance an individual's professionalism. (467)
9. Maintain files and records and prepare reports. (467)

Chapter 20 Answers

Multiple Choice

A.

1. C (481)
2. A (481)
3. C (482)
4. C (483)
5. A (485)
6. C (486)
7. B (485)
8. D (486)
9. A (488)
10. C (489)
11. C (488)
12. D (490)
13. A (498)
14. D (500)
15. C (500)

True/False

B.

1. B The raw data alone cannot support any recommendation. (480)
2. A (480)
3. A (480)
4. A (481)
5. B Records-management systems may be in the following formats: paper or physical, micrographic, electronic. (481)
6. A (480)
7. A (481)
8. B Records can be used in defense of legal challenges in cases of accidents, fatalities, or injuries. (481)
9. B Self-study by an individual should be documented in training records. (482)
10. B Training records may be considered part of an individual's personal, private employment file—a fact that requires an organization to limit access to training records. (482)
11. A (482)
12. A (484)
13. B The widespread use of computers makes the creation of forms very easy. (484)
14. A (488)
15. B The policies, procedures, or guidelines of the training organization need to be continually monitored for effectiveness. (488)
16. B The most common standards used by fire and emergency services organizations in North America are those from NFPA. (491)
17. A (491)
18. B When making the transition to a higher rank, the individual should admit mistakes and errors. (494)
19. B Selection criteria for instructors will be different for short-term versus long-term instructors. (494-495)
20. A (496)
21. B Informal personnel evaluations should be used as a basis for the formal periodic performance review. (497)
22. A (497)
23. A (499)
24. B The training manager or supervisor must continually monitor the job performance of new instructors or staff members and provide appropriate feedback. (499)
25. A (499)

List

C. Types of training documented should include the following items in any order:

1. Daily training delivered by a training division (482)
2. Company training delivered by a company officer or qualified member of the emergency response unit or fire company (482)
3. Organizational training delivered to all members of an organization (482)
4. Self-study by an individual (482)
5. Individual training provided by an organization (482)
6. Special training received outside the organization (482)

Terms

D.

1. Policy—A guiding principle or rule that organizations develop, adopt, and use as a basis or foundation for decision-making (485)
2. Procedure—Identifies the steps that must be taken to fulfill the intent of a policy and is written to support a policy (485)
3. Guideline—Identifies a general philosophy unlike a policy or procedure that provides a clear rule or step-by-step process (486)
4. Adoption—The process by which the chief and/or designated administrators review, amend, and approve the policy, procedure, or guideline (488)

Describe

E. Answer should include points from the following:
The information used in the performance evaluation is gathered from people who have direct professional contact with the person who is being evaluated. The information that is gathered is based on the performance they observe. Responses must remain confidential to protect the people who are providing the information. (500)

Chapter 21 Answers

Multiple Choice

A.

1. D (504)
2. C (504)
3. B (504)
4. D (505)
5. A (505)
6. B (505)
7. C (506)
8. A (509)
9. D (509)
10. C (512)

True/False

B.

1. A (507)
2. A (507)
3. A (507)
4. B While analysis is objective, evaluation is subjective because it depends on the knowledge and experience of the evaluator. (508)
5. B Qualitative evaluation is based most often on nonnumeric analysis. (509)
6. A (511)
7. A (508)
8. B Quantitative evaluation is based on a numeric or statistical analysis. (510)
9. A (510)
10. A (512)
11. A (513)
12. B A formative evaluation is an ongoing, repeated checking during course development and during instruction. (514)
13. B The first step to take after reviewing evaluation results is to determine causes for student failure. (516)
14. A (514)
15. A (515)

Terms

C.

1. Qualitative evaluation—Evaluation that is based most often on nonnumeric analysis (509)
2. Quantitative evaluation—Evaluation that is based on a numeric or statistical analysis (510)
3. Formative evaluation—The ongoing, repeated checking during course development and during instruction to determine the most effective instructional content, methods, aids, and testing techniques (514)
4. Summative evaluation—An end-of-the-course appraisal that commonly measures learning by some form of objective or subjective test (515)
5. Field test—The process of teaching the course on a trial basis (514)

Chapter 22 Answers

Multiple Choice

A.

1. B (522)
2. B (522)
3. A (522)
4. D (523)
5. B (523)
6. C (523)
7. B (524)
8. A (527)
9. A (526)
10. A (534)
11. C (534)
12. B (534)
13. C (538)
14. D (538)
15. C (534)

True/False

B.

1. A (524)
2. A (526)
3. B A goal is NOT an objective because it does not state the performance level of the student. (528)
4. A (530)
5. A (529)
6. B Objectives should be the basis for student testing and evaluation. (530)
7. B Certified Level I or Level II Instructors should teach courses that are specific to the fire and emergency services. (539)
8. A (538)
9. B Evaluating the instructor is considered a part of the evaluation step in the five-step planning model. (540)
10. B A summative evaluation should be performed after every course presentation. (539)
11. A (540)
12. B When planning curriculum/course revisions involve other branches or divisions that may be affected by the proposed changes. (540)
13. A (541)
14. A (540)
15. B Low test scores do NOT always mean that the student did not learn. (537)

Describe

C.

1. Identify—Perform a needs analysis to determine the training program, curriculum, or course required to meet the organization's needs and jurisdictional mandates. (522)
2. Select—Choose the type of training that will meet the requirements. Choose the goals, outcomes, and learning or performance objectives. (522)
3. Design—Design a program, curriculum, or course that will meet the requirements. (523)
4. Implement—Perform a pilot presentation of the course or curriculum. (523)
5. Evaluate—Determine the effectiveness of the course or curriculum in meeting the requirements. (523)